Ecological Guide
to the Mosses & Common Liverworts of the Northeast

Sue Alix Williams

Comstock Publishing Associates
an imprint of
Cornell University Press
Ithaca and London

To my daughter, Arolyn, who began it all

Nan Williams, who struggled with me for many years -
we learned a lot together

and

Jerry Jenkins, without whom this never would have
happened

Copyright © 2023 by Cornell University

All rights reserved. Except for brief quotations in a review, this book, or parts thereof, must not be reproduced in any form without permission in writing from the publisher. For information, address Cornell University Press, Sage House, 512 East State Street, Ithaca, New York 14850. Visit our website at cornellpress.cornell.edu.

First published 2023 by Cornell University Press

Printed in China

Library of Congress Cataloging-in-Publication Data
Names: Williams, Susan Alix, 1957- author.
Title: Ecological guide to the mosses & common liverworts of the
Northeast / Sue Alix Williams.
Other titles: Ecological guide to the mosses and common liverworts of the Northeast
Description: Ithaca : Comstock Publishing Associates, an imprint of Cornell University
Press, 2023. | Includes bibliographical references and index.
Identifiers: LCCN 2022034551 | ISBN 9781501767722 (paperback)
Subjects: LCSH: Mosses--Northeastern States--Identification. |
Liverworts--Northeastern States--Identification.
Classification: LCC QK533.82.N67 W55 2023 | DDC 588/.20974--dc23/
eng/20220809
LC record available at https://lccn.loc.gov/2022034551

Moss

I go outside
and lay in the moss.
Green is all I see.
I look closer. . .
Yes, that's moss.
Even closer. . .
No! Little trees.

I pick it up and look.
Atrichum's leaves curl.
Leucobryum's like a pillow, and
Dicranum's a curly wig.
Hypnum's like a braid, and
Frullania's like a lizard's toe,
Ulota likes to climb up trees,
Polytrichum prefers the ground,
Sphagnum likes the water,
But to me. . .
they all astound

Arolyn Williams
Age 10

Contents

Acknowledgments 6

A Note on Names 6

Author's Note *7*

About This Book *8*

Introduction *9*

- What Is a Moss? 9
- Moss Terminology 10
- Vegetative Reproduction 12
- Acrocarps & Pleurocarps 13
- What Is a Liverwort? 14
- How Is a Leafy Liverwort Different from a Moss? 15
- Incubous vs Succubous 16
- The Well-Equipped Field Bryologist 17
- Helpful Tips 18
- Making a Good Slide 20
- Using the Keys 22

Tree Trunks 23

- Quick & Dirty Field Key 26
- Microscope Key 30
- Photo Gallery 32

Tree Bases 35

- Quick & Dirty Field Key 38
- Microscope Key 44
- Photo Gallery 47

Rotten Logs & Stumps 51

- Quick & Dirty Field Key 54
- Microscope Key 59
- Photo Gallery 61

Humic Soil 63

- Quick & Dirty Field Key 66
- Microscope Key 72
- Photo Gallery 75

Animalcules 78

Mineral Soil 79

Quick & Dirty Field Key 82
Microscope Key 87
Photo Gallery 91

Looking for Gold - Goblins Gold 94

Limy Soil 95

Quick & Dirty Field Key 98
Microscope Key 101
Photo Gallery 103

Dry Rocks & Ledges 105

Quick & Dirty Field Key 108
Microscope Key 115
Photo Gallery 119

Limy Rocks & Ledges 123

Quick & Dirty Field Key 125
Microscope Key 130
Photo Gallery 133

Wet Rocks & Ledges 135

Quick & Dirty Field Key 137
Microscope Key 141
Photo Gallery 143

Wet Soil 145

Quick & Dirty Field Key 148
Microscope Key 155
Photo Gallery 158

Rich Fens & Limy Seeps 161

Quick & Dirty Field Key 163
Microscope Key 166
Photo Gallery 168

Dung Mosses 170

Bryophytes in the Real World 171

A Closer Look at Some Bigger Moss Groups 181

Further Reading 202

Index 203

Acknowledgments

A book like this doesn't just happen in a vacuum. Many, many people throughout the years have helped me along this journey. Howard Crum, Bill Buck, Nancy Slack, Norton Miller and Wilf Schofield have been such an inspiration to me as well as the countless students I've taught over the years both at Eagle Hill and with Jerry Jenkins at his place in White Creek, NY. Also thanks to Glenn Motzkin for giving some helpful suggestions as to the arrangement of some pages as well as more descriptions helpful for beginners.

I am also entirely indebted to Jerry for allowing me to use his stunning photos in this book. It would be a lesser publication without them. All photos are his unless otherwise marked with SAW, which indicate photos I took. His photos are all done using a stacking method and more information as well as many more photos can be found in his book *Mosses of the Northern Forest* as well as on the Northern Forest Atlas's website, https://northernforestatlas.org/.

Over the years I have used many other publications including Crum & Anderson's *Mosses of Eastern North America*, *Flora of North America* vols. 27 & 28, but especially want to give thanks to Bruce Allen's two volumes of *Maine Mosses*. His insights into each species shows an understanding which only someone who has looked at them all has. Finally, *Gathering Moss* by Robin Wall Kimmerer was another book I read many times and one I highly recommend to anyone who has even a passing interest in moss.

A Note on Names

Ah...the stickiness of what to call these mosses. Names keep changing, and what name to use is somewhat of a conundrum. I am writing a book for beginners and will use the names that they will hear and read in the moss world today. The commonly used floras at this moment and recently published field guides use such names as *Ulota crispa* and *Bryhnia graminicolor*. I don't want to muddy the waters by calling them *Ulota crispula* and *Koponeniella graminicolor.* I am following the accepted names published by Tropicos (Missouri Botanic Garden's online taxonomic database) for the moss names and Stotler & Crandall-Stotler's 2017 *A Synopsis of the Liverwort Flora of North America North of Mexico* for accepted Liverwort names. Any names that are newer (or even older but still in use) will be listed in the index. For example, *Ulota crispula* will cross-refer to *Ulota crispa*.

I love bryophytes. I love their complexity and their beauty all wrapped up in such a tiny little package. I love showing people just how cool they are. I've been looking at them for over 25 years, and they never cease to amaze me!

I always ask people how they 'got into mosses' because it's usually a convoluted and often interesting trip. Here's my story:

It was 1993 and I was working on a project for the state on the rare orchid *Triphora trianthophora*. My eldest daughter, Arolyn, at the time was 8 and quite interested in nature and being outside. After one trip with a very talented naturalist whose specialty was mushrooms, Arolyn decided that SHE wanted a specialty and informed me that it was going to be mosses. Well, at that time, I knew absolutely nothing about mosses. They all looked the same to me. So, my mother-in-law, Nan, and I identified about 16 common mosses and helped Arolyn to learn them too. At some point during the summer, the state botanist came up to check on the orchid project and Arolyn was tagging along and telling him, "Oh look, there's haircap moss," or "there's *Dicranum flagellare*." He was very impressed and said she should go to Eagle Hill for the moss seminar that Howard Crum taught. I found out it was usually for professionals and sort of laughed and said maybe when you're older. We returned home to NH and Arolyn was still all gung ho, so I told her to write a letter to Eagle Hill, which she promptly did, naming all the mosses that she knew. I figured that was the end of that. Well, to my surprise, sometime later, I got a call from Joerge at Eagle Hill, and he said he talked to Howard Crum who said that Arolyn could come to the weeklong seminar, for free! This was the weekend before it was to take place, and Nan was at our house at the time and she said, "Go, I'll take care of Meredith (my younger daughter), just go." So we did. It was an amazing time. Howard Crum was so kind; he would often say, "Oh, here's a moss that Arolyn knows," and she would pipe up in her little voice and say, "*Dicranum flagellare*." All the other adults were amazed that she could remember the Latin names. The week was overwhelming with so much information. Like all children, Arolyn moved on to different things, but I stuck with those mosses from that time on. The trip hasn't been an easy one. Learning mosses is hard, and harder still without a field guide that a beginner can use.

I looked at a LOT of moss that I never could identify, but I learned something from each of them. I started drawing mosses because it helped me to identify them, I had to look, really look, at a specimen to draw it and that helped me to use the keys better. There are still days when I collect something, and I can't figure out what it is. Mosses like to trick you.

This is the guide I wish I had had when I was starting out, and I'm hoping that it will help you to appreciate the miniature marvels that bryophytes are with much less frustration than I had at the beginning!

See you in the field!

Sue

If you have flipped through this book, you will have noticed that it's not set up like other keys or guides to mosses or plants. You won't find any dichotomous keys in the beginning of each section. Instead, I have opted to segregate the mosses by substrate. The problems most beginners face are two-fold: (1) not understanding the terminology used, therefore, using a key becomes a ponderous task of looking back and forth between the key and the glossary (if there is one available), and (2) not understanding what the key is asking them to look at (or not having the equipment, e.g., a compound microscope). What does it mean, "Is it papillose?" Therefore, they have to look through every single option. This quickly turns into just flipping through the pictures and becoming discouraged.

The ONLY sure thing I knew when I first started looking at moss was what it was growing on. I didn't know if its cells were papillose (I didn't even know what that meant). I wasn't sure where exactly the alar cells were...but I knew that I found it on a log. Only after many years of looking at bryophytes in the field was I able to understand that I didn't HAVE to look at every single moss in the book. *Fontinalis* wasn't going to be on a tree trunk and *Ulota crispa* wasn't going to be on the ground. Now when I go in the field searching for bryophytes, I automatically narrow the list down in my head to the possibilities I'm LIKELY to find on that substrate. If I'm in a second growth woods, I expect to see *Ulota crispa*, *Platygyrium repens* and *Frullania eboracensis* on tree trunks. If I'm in an old growth forest, I'll start looking for *Neckera pennata*, *Leucodon andrewsianus* and *Haplohymenium triste*. If I'm looking at rocks in a brook, I'll be expecting *Scapania*, *Hygrohypnum* and *Torrentaria riparioides*.

This is what I've tried to do for you in this book. I've tried to narrow down the possibilities of what you might actually find on a particular substrate.

A NOTE OF WARNING. Anyone who is familiar with bryophytes will tell you that they are not as faithful to substrates as this book implies - and they are right. I have done my best to put the MOST COMMON or unique species into the area where I find them most frequently. A few species are generalists and are equally common on a variety of substrates. I have included those in each appropriate section. Others, while most common in one or two substrates, occasionally like to try a new one. This is the nature of bryophytes. If you don't find what you're looking for, try the next, most likely substrate. Sometimes there's not much difference between a log and soil and a tree base. Or for mosses in the section of Rich Fens & Limy Seeps, you might have to go into the Wet Soil section.

What Is a Moss?

Mosses are small, green, spore-bearing plants. Most of them have a stem with leaves on it similar to 'higher' plants, but their leaves are usually only one cell thick. They have things that look like little roots, called rhizoids, which are used for anchoring the plant to its substrate. They have no true vascular tissue, and all water and nutrient uptake is done mainly through the outer leaf walls. They have also evolved many interesting structures that help in the quick hydration of the plant as well as maintaining hydration for longer times. Many mosses have the amazing ability to dry out completely, suspending all growth and then, when water becomes available again, resume photosynthetic metabolism. Owing to their small size and seeming insignificance, many people assume they are a 'lower' plant form. However, they are very specialized and an extremely successful group of plants that occupy almost every possible niche available to them. Their life cycle, which is similar to liverworts and hornworts, alternates between a leafy sexual generation and a non-leafy spore bearing one.

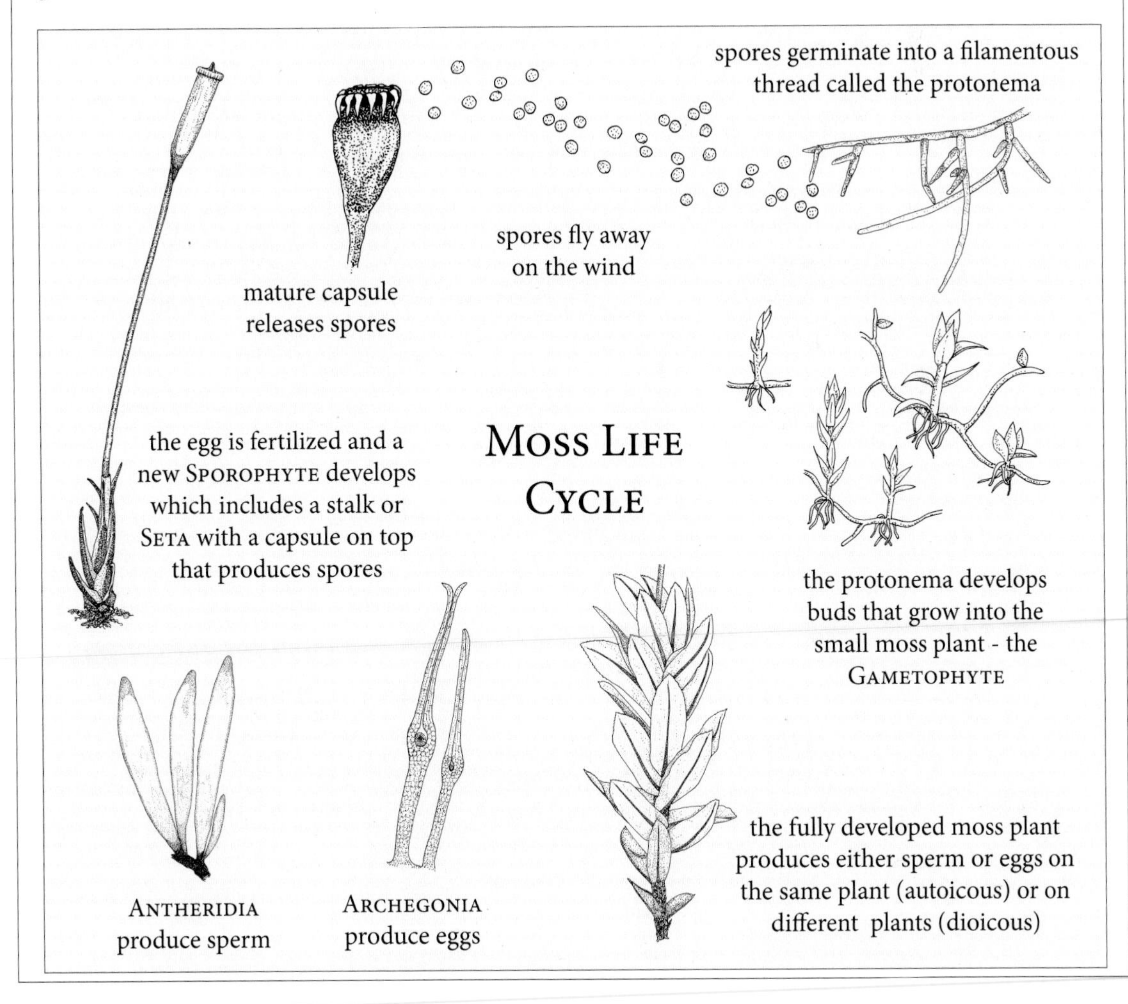

MOSS TERMINOLOGY

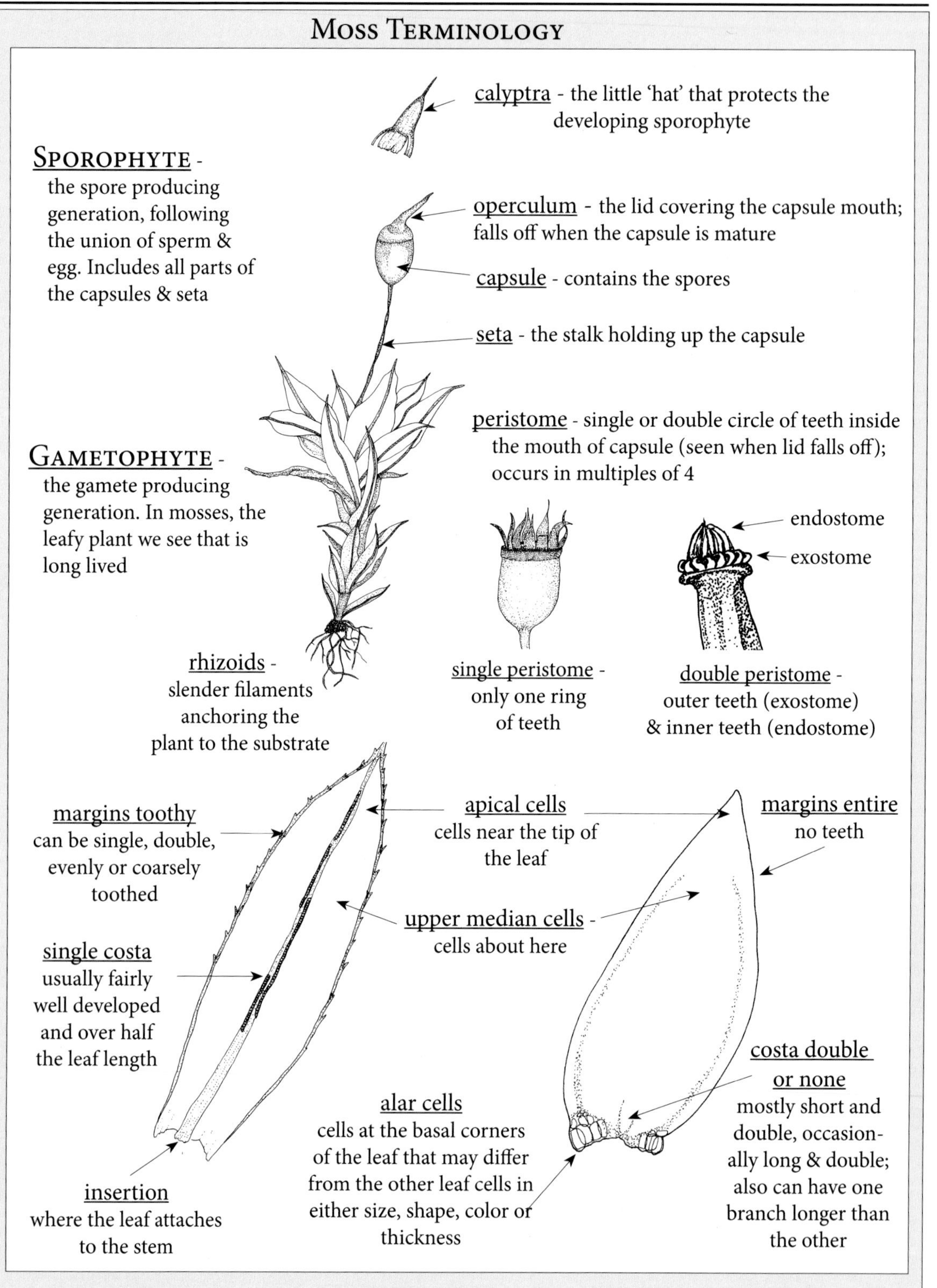

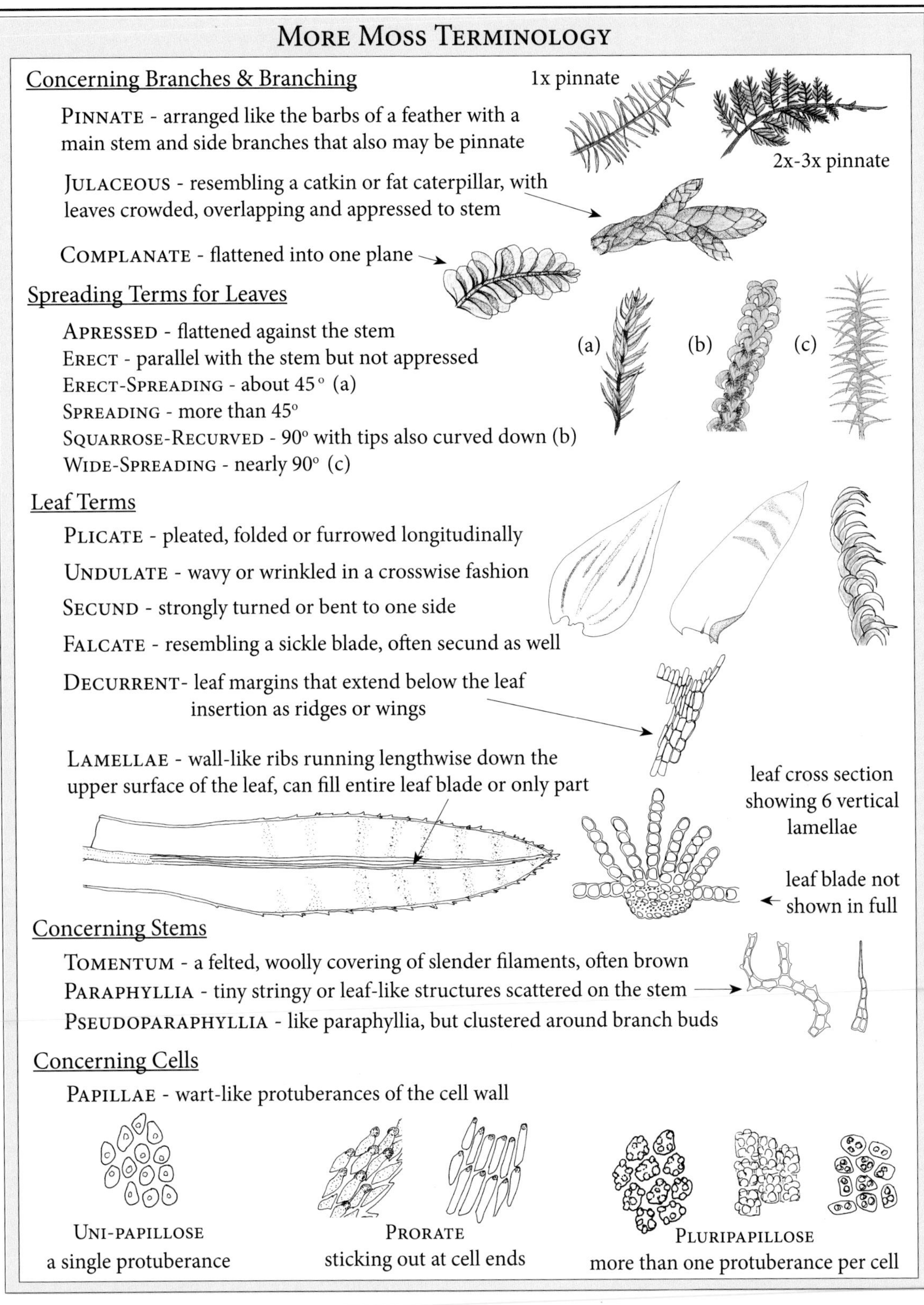
More Moss Terminology
Concerning Branches & Branching
1x pinnate
2x-3x pinnate
Pinnate - arranged like the barbs of a feather with a main stem and side branches that also may be pinnate
Julaceous - resembling a catkin or fat caterpillar, with leaves crowded, overlapping and appressed to stem
Complanate - flattened into one plane
Spreading Terms for Leaves
Apressed - flattened against the stem
Erect - parallel with the stem but not appressed
Erect-Spreading - about 45° (a)
Spreading - more than 45°
Squarrose-Recurved - 90° with tips also curved down (b)
Wide-Spreading - nearly 90° (c)
(a)
(b)
(c)
Leaf Terms
Plicate - pleated, folded or furrowed longitudinally
Undulate - wavy or wrinkled in a crosswise fashion
Secund - strongly turned or bent to one side
Falcate - resembling a sickle blade, often secund as well
Decurrent- leaf margins that extend below the leaf insertion as ridges or wings
Lamellae - wall-like ribs running lengthwise down the upper surface of the leaf, can fill entire leaf blade or only part
leaf cross section showing 6 vertical lamellae
leaf blade not shown in full
Concerning Stems
Tomentum - a felted, woolly covering of slender filaments, often brown
Paraphyllia - tiny stringy or leaf-like structures scattered on the stem
Pseudoparaphyllia - like paraphyllia, but clustered around branch buds
Concerning Cells
Papillae - wart-like protuberances of the cell wall
Uni-papillose
a single protuberance
Prorate
sticking out at cell ends
Pluripapillose
more than one protuberance per cell

VEGETATIVE REPRODUCTION

When it comes to making new plants, mosses often like to hedge their bets. Producing reproductive structures is costly to the plant and takes a long time, sometimes almost a whole year before the spores are ready. So, many mosses (and liverworts) have vegetative propagules that are less costly energetically to the plant and can be produced much quicker to colonize unoccupied substrates nearby quickly.

These propagules can be in an undifferentiated form of few to many cells called gemmae. Gemmae usually will grow into a protonematal mat from which the new plants form. Brood bodies or brood branches are more differentiated...they often look like little branches, some even with reduced leaves. These can grow directly into a new plant or into protonema.

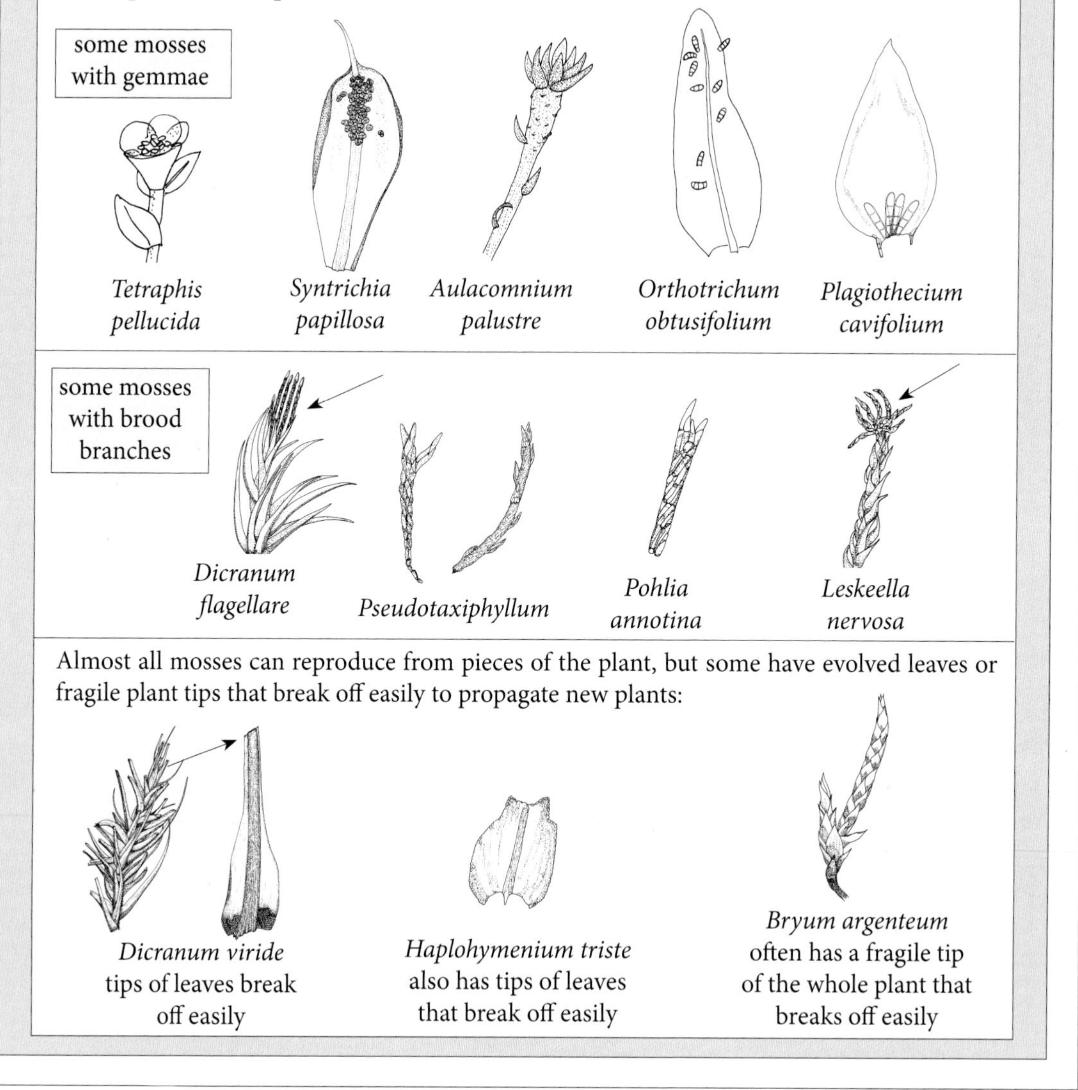

Mosses are usually divided into two large groupings:

ACROCARPS & PLEUROCARPS

ACROCARPS have a main stem that generally grows upright. They are simply branched or sometimes forked.

The sporophytes (with the capsule) arise from the tip of the stem or main branch.

Most of the Acrocarps in our area have leaves with a well developed midrib (costa).

Growth forms include:

Cushions - dense, rounded tufts of crowded and radiating stems

Turfs - large patches of short or tall, crowded, erect shoots

Groves - small to large groupings of plants not crowded together

PLEUROCARPS have a main stem that grows mostly prostrate with many branches. These branches can be regular as in pinnate branching or, more often, irregular. Sporophytes arise from a side branch somewhere below the tip of the main branch.

Growth forms include:

Mats - stems are flattened to the substrate and usually adhere rather tightly

Stringy Mats - small mosses, loosely attached and often hanging down from substrate like a tree trunk or rock

Big & Messy - large, irregularly branched, loose mats; called wefts by some

Shelves or Sprays - sticking out away from a vertical substrate; can be pinnate and overlapping or hooklike

NOTE: MOSSES ENLARGED TO SHOW DETAIL!

WHAT IS A LIVERWORT?

Liverworts are also small, green plants grouped together with mosses within the Bryophyte group because of their similar life cycles. Liverworts are of three different types: SIMPLE THALLOID consists of solid tissue and lacks pores and scales, COMPLEX THALLOID consists of loose upper tissue and generally has dorsal pores and ventral scales and LEAFY have leaves on a stem and are the ones most likely to be confused with mosses.

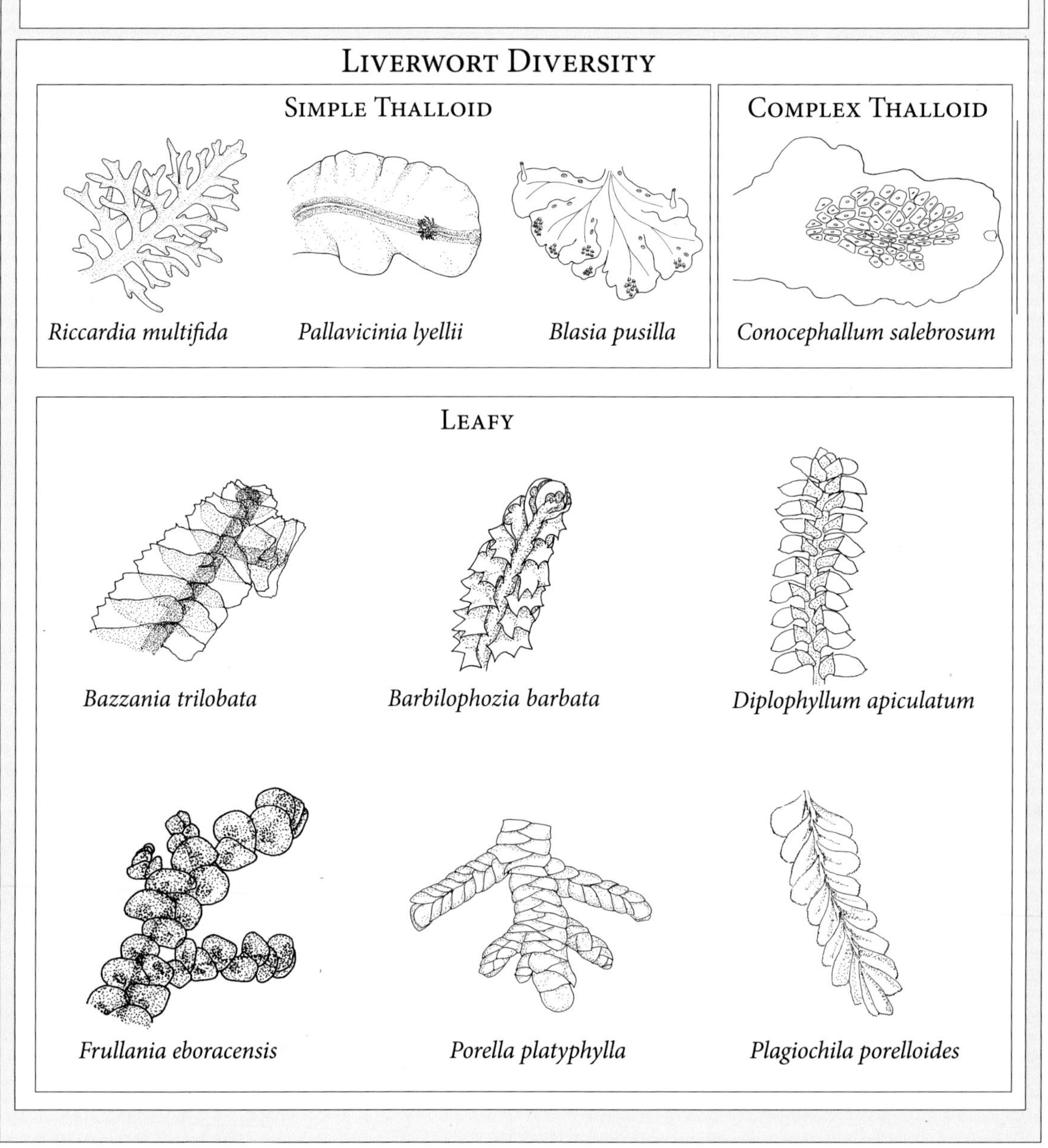

How Is a Leafy Liverwort Different from a Moss?

Liverwort

Leaves inserted in 2 rows, sometimes with a 3rd row of smaller or similar underleaves

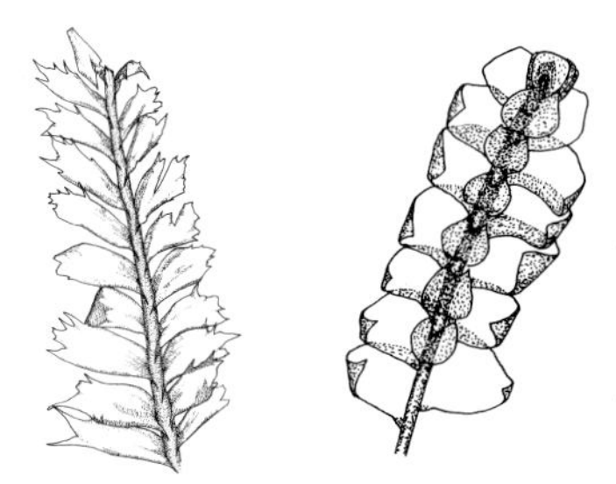

Leaves often lobed and without a midrib

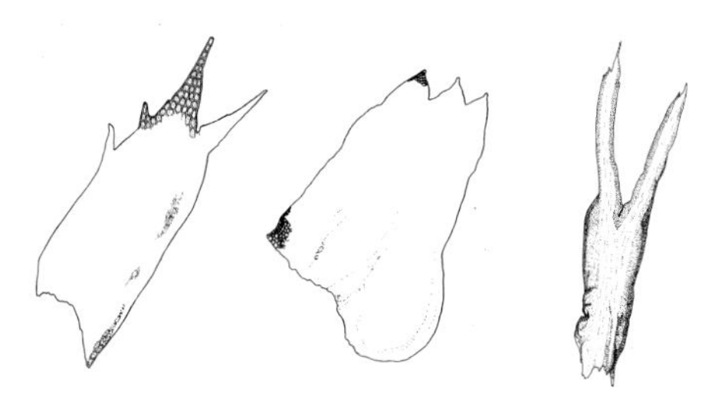

Seta colorless, fragile, quickly disintegrating. Capsules open by longitudinal lines and never have teeth around a mouth. Capsules are quite similar for most liverworts.

Moss

Leaves usually inserted in more than 3 rows

Leaves never lobed and often have a midrib

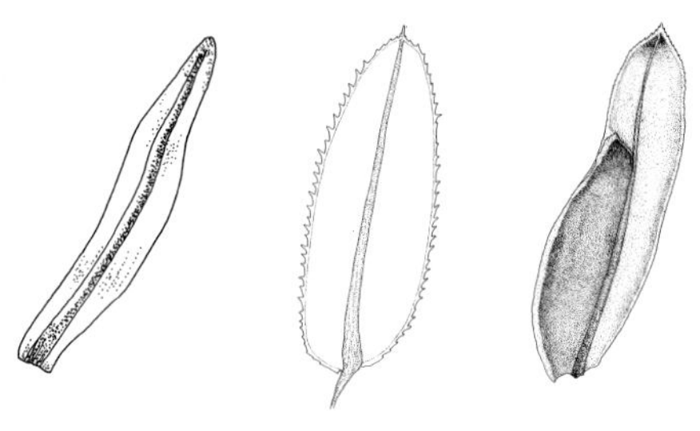

Seta colored, wiry and persistent. Capsules mostly open by a lid and often have teeth around mouth. Capsule shape and size quite variable between genera and species.

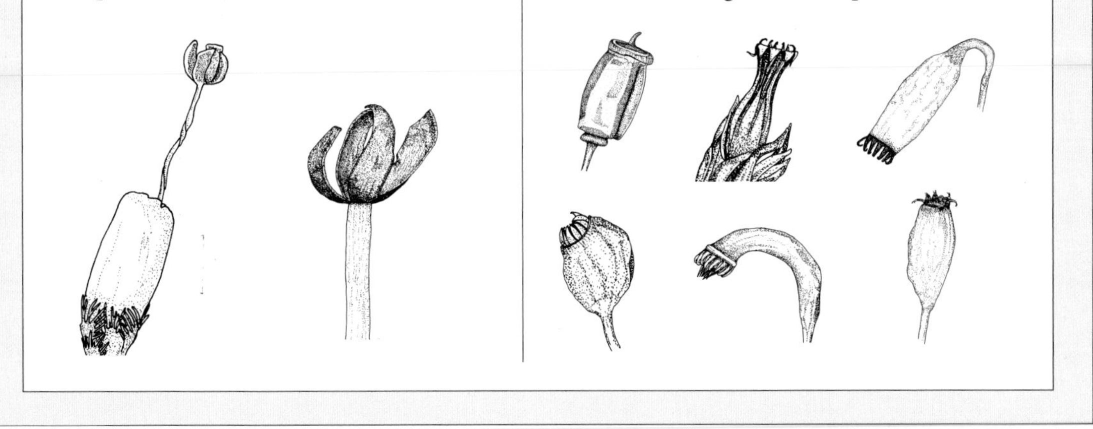

INCUBOUS VS SUCCUBOUS

In studying liverworts you will run across the terms INCUBOUS and SUCCUBOUS when referring to how some leafy liverwort leaves are inserted on the stem, but what does that exactly mean and how can you remember which is which? I've got just the way for you!

Suppose we decided to make two huts - one with a roof of *Calypogeia* (leaves incubously inserted) and one of *Barbilophozia* (leaves succubously inserted).

(Note that the growing tips are at the top of the roof)

Calypogeia hut

INcubous leaves - let the rain IN

Incubous leaves are inserted so you can see the top edge of the leaf

Barbilophozia hut

Succubous leaves - don't let the rain in

Succubous leaves are inserted so you can see the bottom edge of the leaf. They are how roof shingles are put on

THE WELL-EQUIPPED FIELD BRYOLOGIST

Handlens - This is an essential item for looking at bryophytes! Use 7x or 10x for general purposes. Hang from a lanyard around your neck.

Pocket knife - Not essential, but nice to have. Makes collecting mosses from rocks and bark easier. Brightly colored is good for when you leave it on the ground (which you will!).

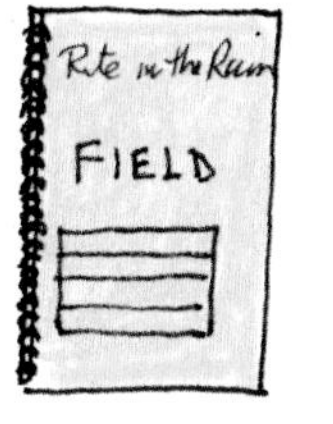

Field Notebook - Not essential, but it's nice to keep notes on where you found your specimens. You can include a lot of information in here. There are notebooks that you can use when the weather is bad and they don't turn to mush (Rite-in-the-Rain).

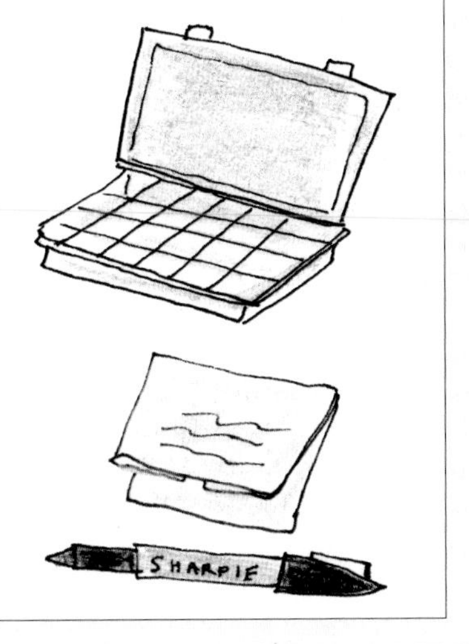

You can collect your bryophytes in a plastic box, paper packets or small paper bags.

The box is convenient, holds samples nicely, ensures you collect small samples and are easy to get to. HOWEVER, you must not leave the box closed for days, or your specimens will get all moldy. Paper packets require you to make them beforehand, but they are absorbant (especially if made of newspaper), you can just leave the specimens in them until you're ready to look at them, and you can write information directly on the packet in the field.

HELPFUL TIPS

DO: Flip through this book and look at all the pretty pictures and notice, contrary to popular belief, that all mosses do not look the same!

DO: Go out for a walk without your handlens and just start noticing the mosses. Look at the colors and textures and see if you can tell that some of them are different from others…look on rocks and trees and on the ground…you don't have to go anywhere special, they grow everywhere.

DO: Try to pick out a moss that looks really distinctive to you, keep walking and see if you can pick it out again. I call this the 'there it is again' technique, and this will be critical in making your first successful identifications.

DO: Describe the moss out loud as you're looking at it, starting with the capsule. Is it erect or curved? Does it have a teeth? Long seta or none? What shape are the leaves? Are they toothed? Does it have a costa? Any vegetative (gemmae) visible? The more you can describe it, the more you have to work with to try to identify it.

DON'T: Collect the scrawniest, scrappiest moss you can find thinking it's going to be something rare because usually it's just a scrawny, scrappy piece of something common that just makes it hard, or usually impossible, to identify.

DON'T: Automatically assume that the capsules you see go with the moss that you're looking at! Mosses like to hang out together and grow all mixed up...that capsule might be coming from a moss growing underneath or mixed in the one that you are trying to identify.

DON'T: When collecting mosses to bring home to identify - don't take the entire patch! Bryophytes take a long time to grow and you need only a tiny piece for identification. Take the smallest amount possible for your needs in determining that moss.

WRONG: Lens close to plant and far from eye, both eyes open

RIGHT: Lens close to eye, bringing plant up to you

OR: 'Assume the position' head down, butt up!

MAKING A GOOD SLIDE

In order to identify mosses, we often need to look at certain features under a compound scope: whether it has a costa or not...the shape of the cells...does it have differentiated alar cells...are the cells papillose? We need to prepare our specimen properly in order to see clearly what it is we want to see.

PRACTICE, PRACTICE, PRACTICE!

1

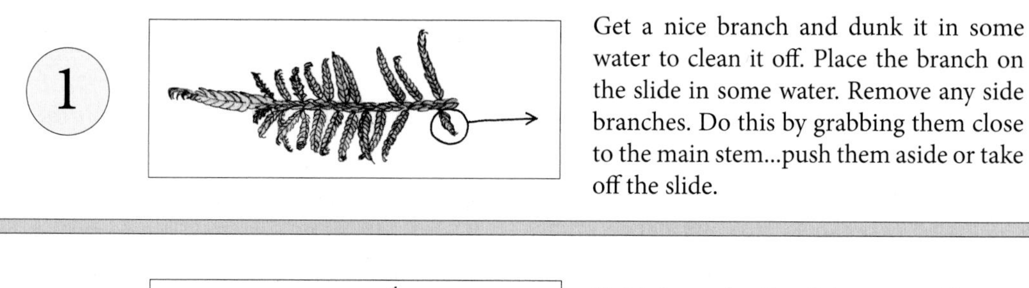

Get a nice branch and dunk it in some water to clean it off. Place the branch on the slide in some water. Remove any side branches. Do this by grabbing them close to the main stem...push them aside or take off the slide.

2

Hold down the tip of the stem with your needle tool. With your tweezers, hold the stem on either side fairly close to the tip.

3

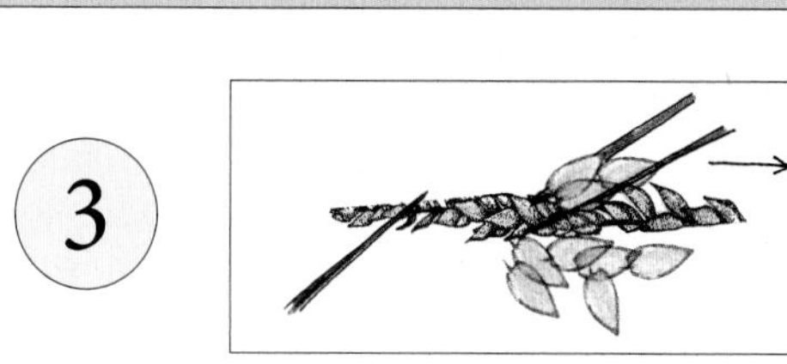

With the tweezer tips on either side of the stem (and right up against it), VERY GENTLY pull the tweezers towards the base of the stem, stripping the leaves off the main stem.

4

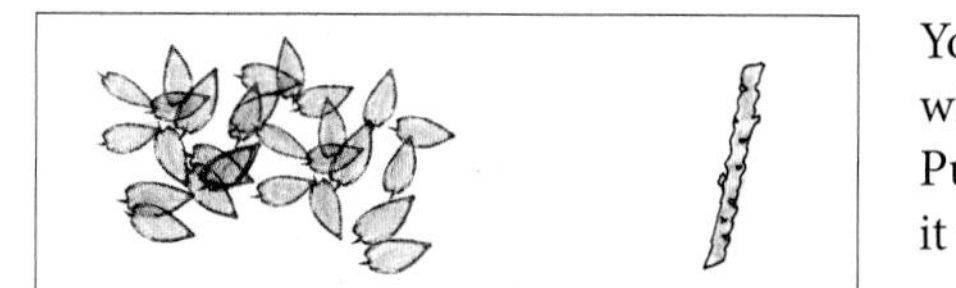

You'll now have a hopefully naked stem with a pile of leaves in a puddle of water. Put the stem to the side (you might need it to make cross sections later).

5

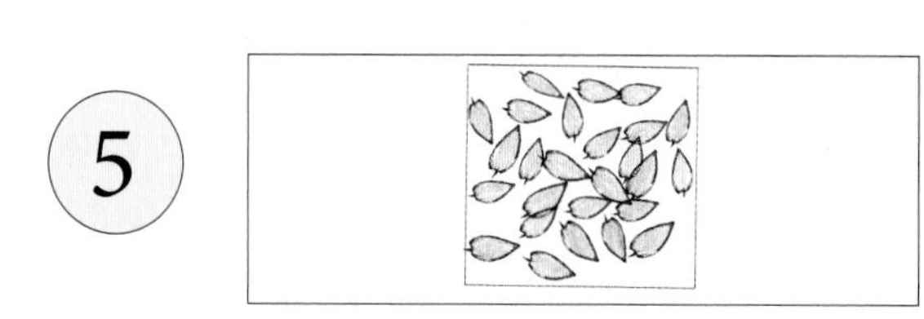

Push the leaves around in the water with the needle tool so they are not overlapping and piled up. You want a LOT of leaves on your slide, not just one or two. What you will be looking at is basically the AVERAGE leaf shape, costa, etc. Put a cover slip on and you're ready to go!

Sometimes, in order to determine a species, you really need to make cross sections of leaves or stems. You might want to tell sterile Polytrichums apart for example or to tell *Hygrohypnum eugyrium* from *Sematophyllum marylandicum*. This is the technique I use to make sections of leaves as well as stems.

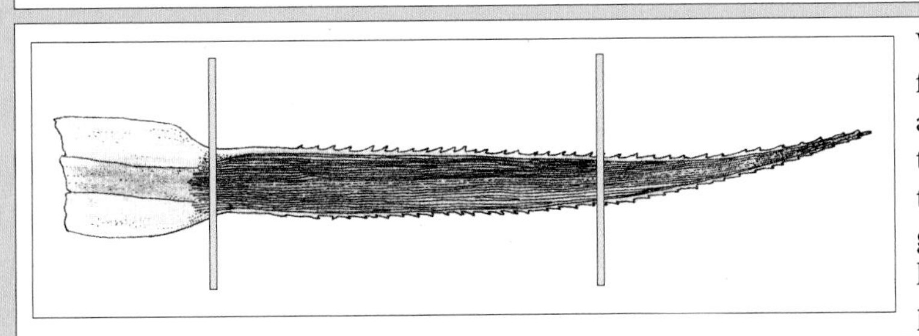

With Polytrichums, I take a few leaves and cut off the tips and bases of the leaf to make the leaves lie flat and give me the best section from which to get lamellae. It's easiest if the leaves are moist but not sopping wet.

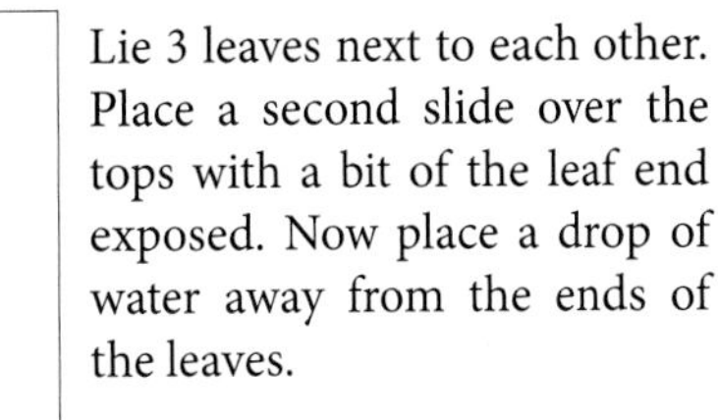

Lie 3 leaves next to each other. Place a second slide over the tops with a bit of the leaf end exposed. Now place a drop of water away from the ends of the leaves.

side view

Place your finger a bit over the end of the top slide (this will be your guide). With a very sharp razor blade, cut vertically through the leaf tips. Try to make the slimmest cut possible....Then....

side view

Tip the blade to about a 45 degree angle and drag the cut pieces into the drop of water. Lift the blade out of the water and go back to making more cuts. Make a LOT! If the drop of water ends up under the slides, the task is much more difficult.

Move the top slide back as needed to expose more leaf to cut. After cutting many cross sections, pull the top slide back quite far exposing a larger section of leaf. With the blade at a 45 degree angle, gently scrape off the lamellae and drag to the pool of water. Don't push too hard or you'll cut the leaf in half. Now you have a pool of water with both lamellae cross sections and lamellae lying sideways.

USING THE KEYS

USING THE QUICK & DIRTY FIELD KEYS

Yellow boxes tell you something about the growth habit of certain species.

+ Plus signs indicate the characters I use to identify a particular species in the field. Some don't need many characters, some need quite a few. If you don't have all the characters, and there are other species in that group, you'll need to go to the MICRO key to sort them out.

BOLD - indicates a common species you will most like encounter
REGULAR - indicates an uncommon or infrequent species
RED - indicates a rare species

A magnifying glass shows that you can identify this species in the field if you have all the necessary characters.

A microscope means you need a scope (dissecting and/or compound) in order either to confirm a field ID or to sort out some species that can only be confirmed by microscopic characters.

USING THE MICROSCOPE KEYS

These sections gives more detailed information about the structure of a specific moss. If you have only a dissecting microscope, you can still determine if a leaf is toothy or not, or if it has a midrib (costa), and sometimes you can see differentiated alar regions as well as capsule structure.

The MICRO KEY is broken down into groupings, Acrocarps/Pleurocarps, Costa/No costa, Toothed/ Untoothed, as well as showing details important to IDing, such as any gemmae, or cell size and papillosity.

The **PHOTO GALLERY** for each section shows some of the species from that section, not every one.

The **BRYOPHYTES IN THE REAL WORLD** section shows actual places and what common or unusual species are found there and on which substrates. A variety of places are illustrated to give an idea of what you might find.

Last, the section on **MOSS GROUPS** looks closer at some of the bigger and/or more difficult groups encountered. This section does not cover every genus, and some are lumped together. Most of the difficulties can be figured out from the sections on each substrate.

NOTE: I have included only the barest information concerning liverworts in this book. Only the largest, most commonly found or unique are included, and they are shown only in the Field Keys and sometimes in the Photo Gallery.

TREE TRUNKS

What better place to start looking at mosses than on tree trunks. They're right at eye level - no bending over for these! Mosses included in this section of the book are those found higher than 2 1/2 to 3 feet above the ground; any further down and you're getting into the nebulous tree base region! Both mosses and liverworts are found on trees, and the nice thing is that, in our area, the bryoflora is limited and predictable.

How many bryophytes will you find? Perhaps 16 or so mosses with 10 common ones and about 5 liverworts, 4 of which will be fairly common.

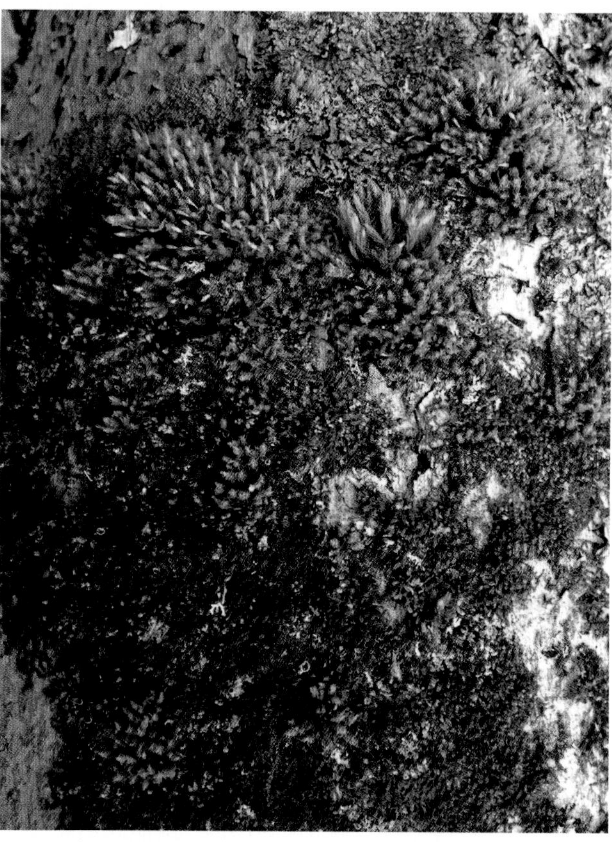

The BIG 3 bryophytes of tree trunks are: *Ulota crispa*, *Platygyrium repens* and *Frullania eboracensis*. These three are so common, that if you learn them, you will know what grows on tree trunks (in our area) about 75% of the time!

There are several different growth forms that you will notice:
Cushion formers - these are many small upright plants crowded together to form either dense or loose cushions. Examples of these include *Ulota crispa*, *Dicranum montanum* and *D. viride* and the Orthotrichums.

Mat formers - these mosses form low, dense mats tight to the trunk and include such bryophytes as *Platygyrium*, *Hypnum pallescens*, *Frullania* and *Radula*. In moister areas or on large, old trees you get a looser, deeper carpet, such as *Anomodon attenuatus*.

Finally you have the **Shelf-like form**, seen in such bryophytes as *Neckera pennata*, *Leucodon andrewsianus* and the liverwort *Porella platyphylla*, where the plant grows in overlapping shelves and sticks out away from the trunk of the tree.

There is also a difference between which species you might find on a tree depending on the type of forest you are in and what kind of tree you're looking at. Species on apple trees in an orchard will be different from those found in rich beech/maple woods or drier oak/hickory woods or an old willow next to a stream. Trees found in old growth woods will also have a different community of bryophytes.

A Sampling of Tree Trunk Mosses

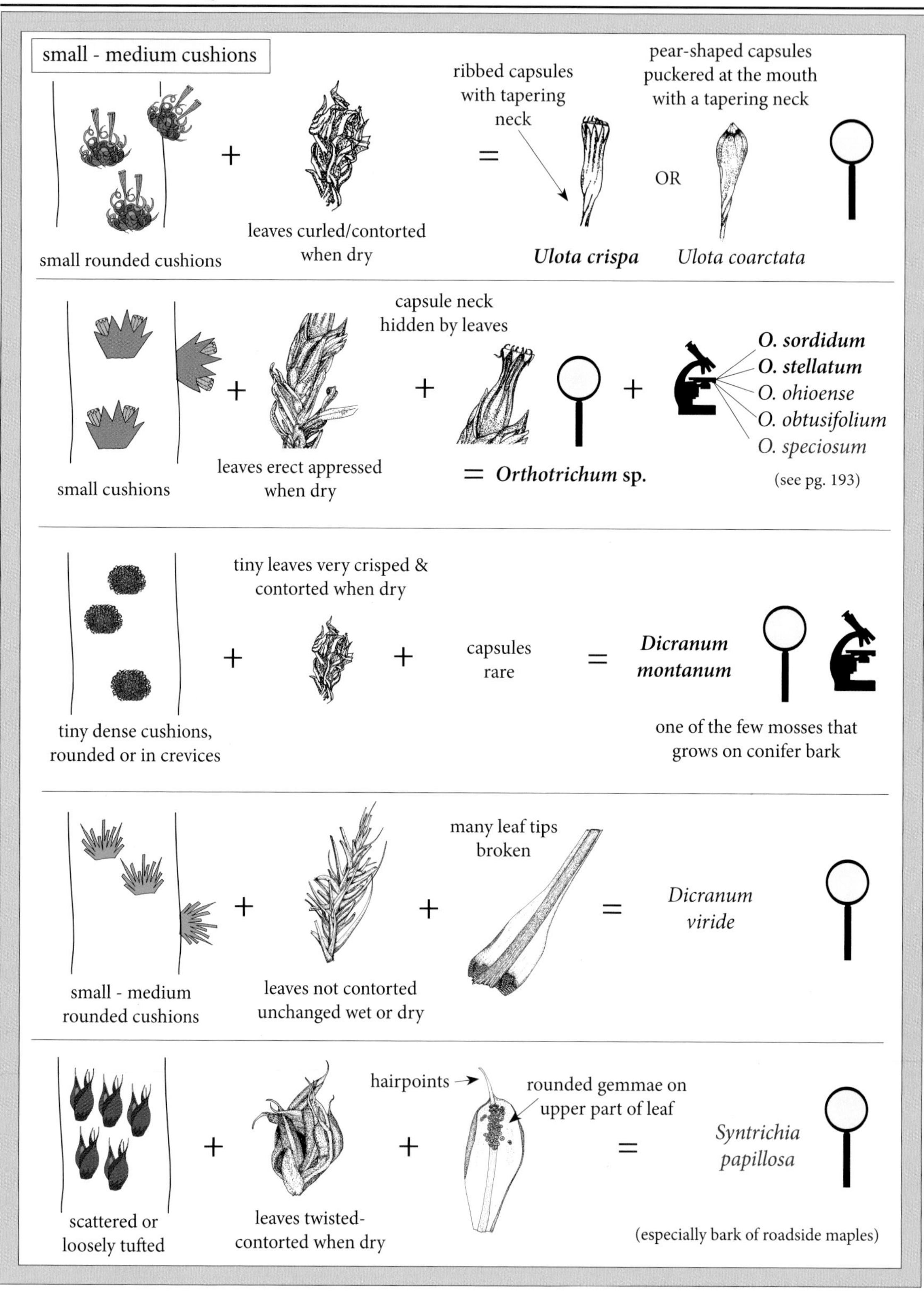
small - medium cushions
small rounded cushions
+
leaves curled/contorted when dry
=
ribbed capsules with tapering neck
OR
pear-shaped capsules puckered at the mouth with a tapering neck
Ulota crispa
Ulota coarctata
small cushions
+
leaves erect appressed when dry
+
capsule neck hidden by leaves
+
O. sordidum
O. stellatum
O. ohioense
O. obtusifolium
O. speciosum
(see pg. 193)
= Orthotrichum sp.
tiny dense cushions, rounded or in crevices
+
tiny leaves very crisped & contorted when dry
+
capsules rare
=
Dicranum montanum
one of the few mosses that grows on conifer bark
small - medium rounded cushions
+
leaves not contorted unchanged wet or dry
+
many leaf tips broken
=
Dicranum viride
scattered or loosely tufted
+
leaves twisted-contorted when dry
+
hairpoints
rounded gemmae on upper part of leaf
=
Syntrichia papillosa
(especially bark of roadside maples)

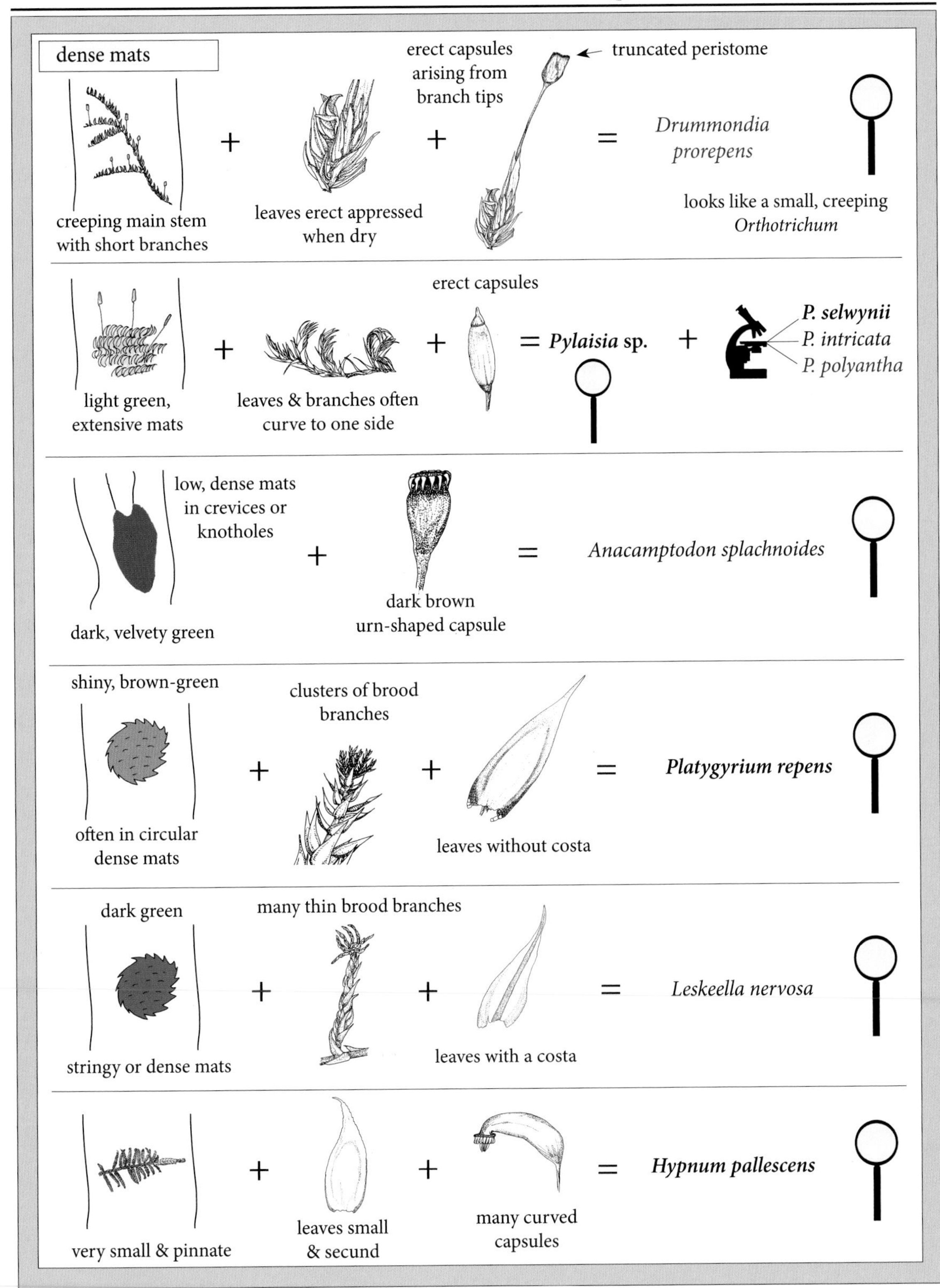
dense mats
erect capsules arising from branch tips
truncated peristome
creeping main stem with short branches
leaves erect appressed when dry
Drummondia prorepens
looks like a small, creeping Orthotrichum
erect capsules
light green, extensive mats
leaves & branches often curve to one side
Pylaisia sp.
P. selwynii
P. intricata
P. polyantha
low, dense mats in crevices or knotholes
dark, velvety green
dark brown urn-shaped capsule
Anacamptodon splachnoides
shiny, brown-green
often in circular dense mats
clusters of brood branches
leaves without costa
Platygyrium repens
dark green
stringy or dense mats
many thin brood branches
leaves with a costa
Leskeella nervosa
very small & pinnate
leaves small & secund
many curved capsules
Hypnum pallescens

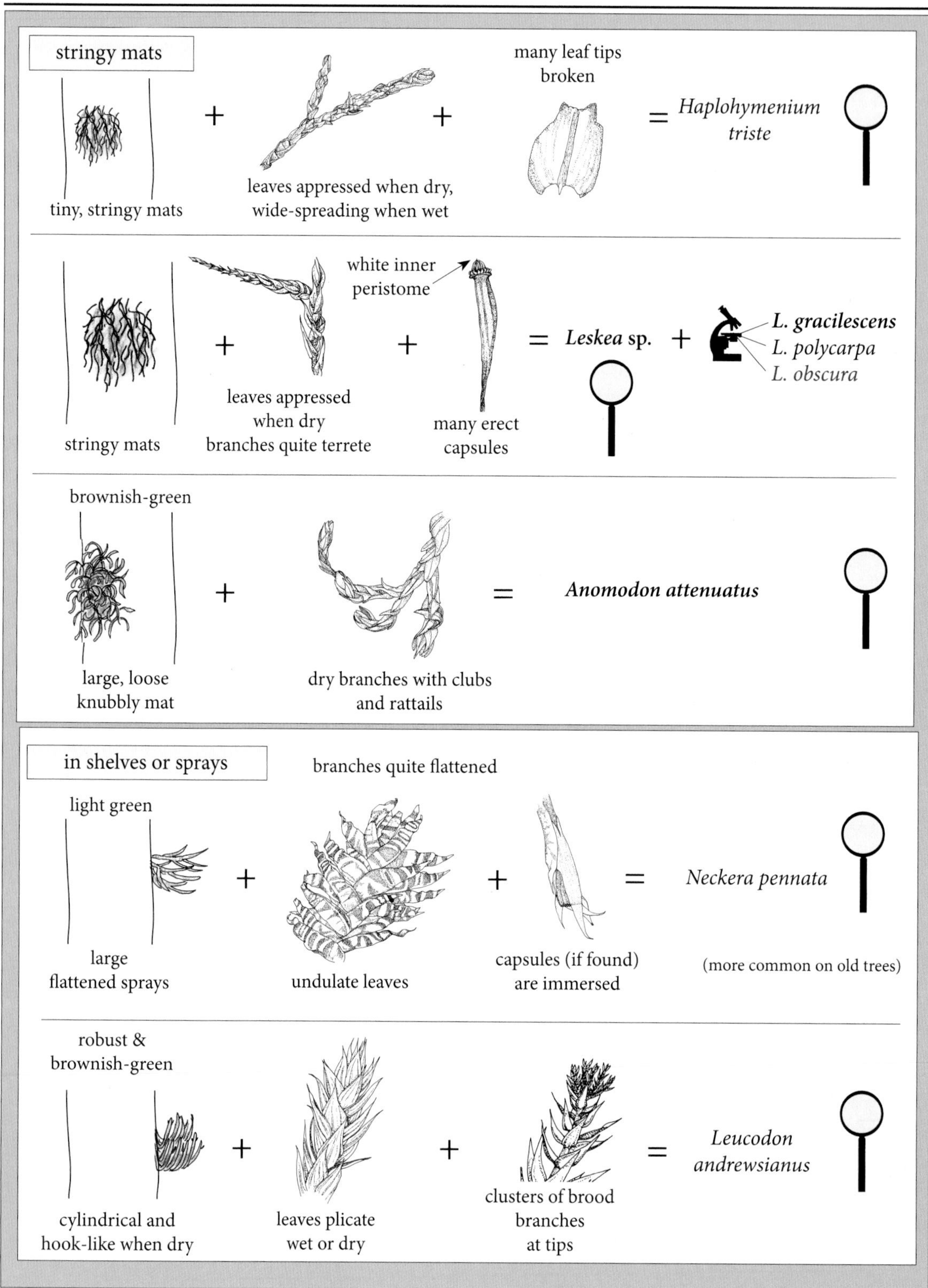
stringy mats
tiny, stringy mats
+
leaves appressed when dry, wide-spreading when wet
+
many leaf tips broken
= Haplohymenium triste
stringy mats
+
leaves appressed when dry branches quite terrete
+
white inner peristome
many erect capsules
= Leskea sp.
+
L. gracilescens
L. polycarpa
L. obscura
brownish-green
large, loose knubbly mat
+
dry branches with clubs and rattails
= Anomodon attenuatus
in shelves or sprays
branches quite flattened
light green
large flattened sprays
+
undulate leaves
+
capsules (if found) are immersed
= Neckera pennata
(more common on old trees)
robust & brownish-green
cylindrical and hook-like when dry
+
leaves plicate wet or dry
+
clusters of brood branches at tips
= Leucodon andrewsianus

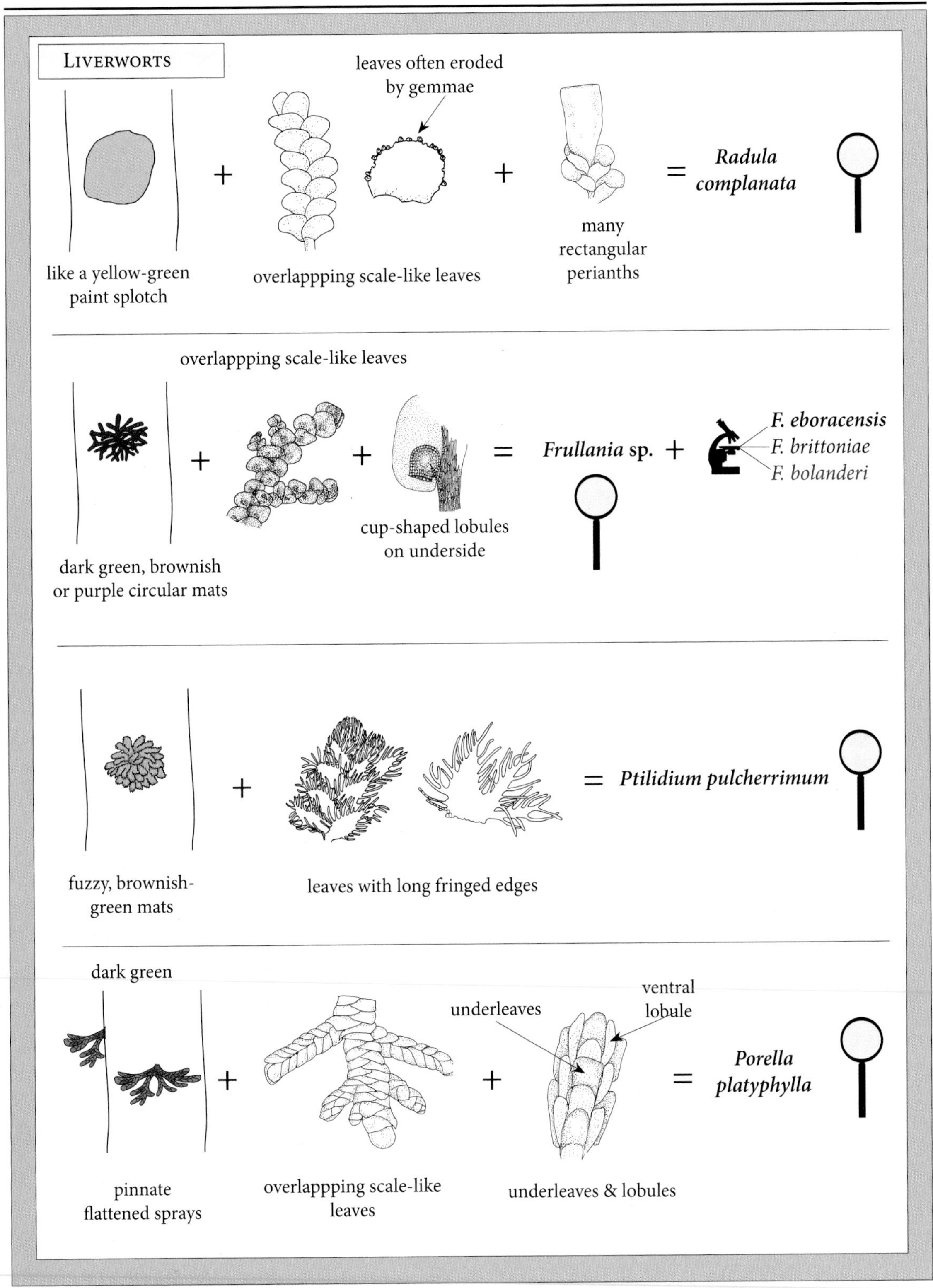
LIVERWORTS
leaves often eroded by gemmae
+
+
=
Radula complanata
like a yellow-green paint splotch
overlappping scale-like leaves
many rectangular perianths
overlappping scale-like leaves
+
+
=
Frullania sp. +
F. eboracensis
F. brittoniae
F. bolanderi
dark green, brownish or purple circular mats
cup-shaped lobules on underside
+
=
Ptilidium pulcherrimum
fuzzy, brownish-green mats
leaves with long fringed edges
dark green
underleaves
ventral lobule
+
+
=
Porella platyphylla
pinnate flattened sprays
overlappping scale-like leaves
underleaves & lobules

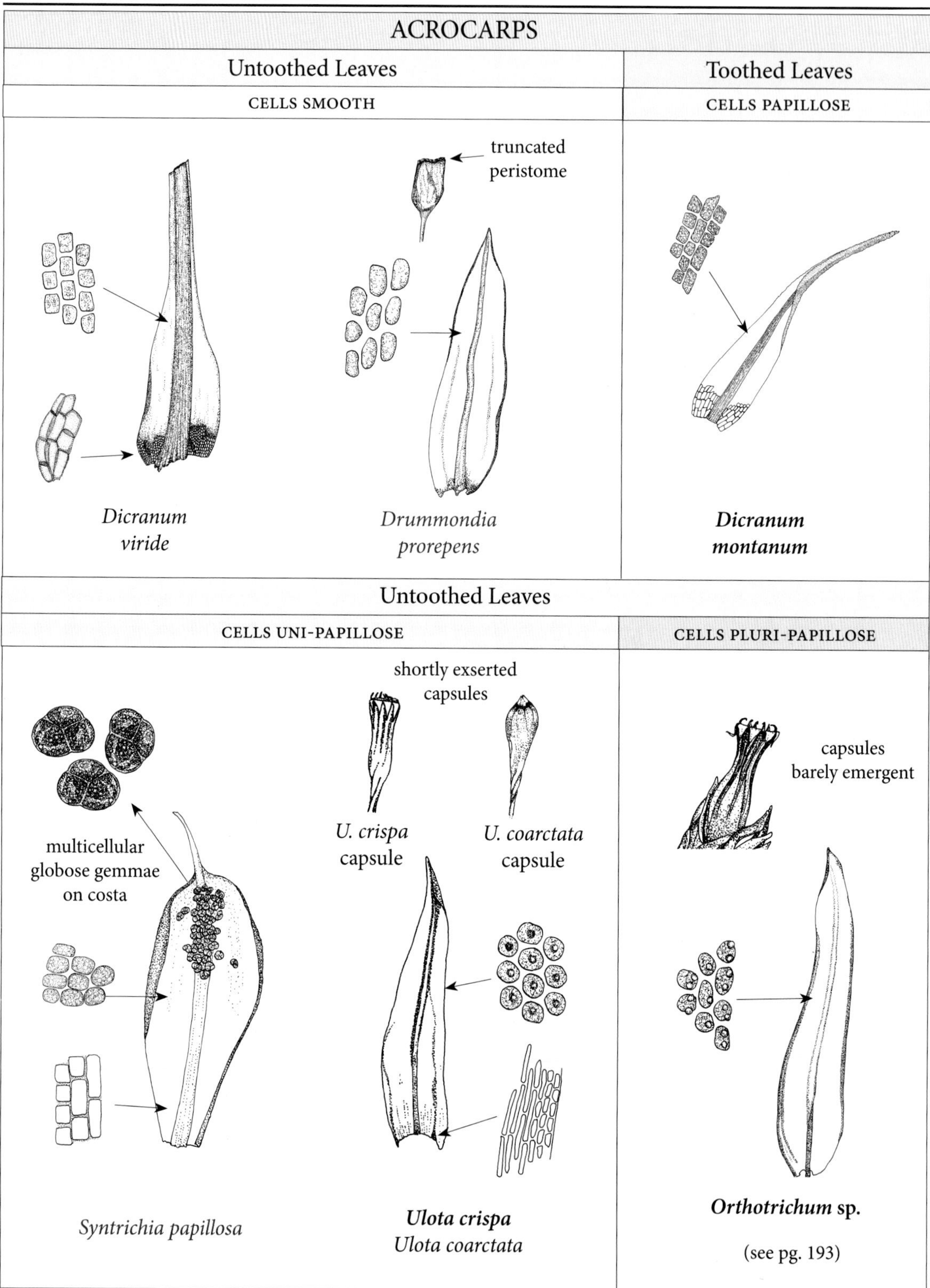
ACROCARPS
Untoothed Leaves
Toothed Leaves
CELLS SMOOTH
CELLS PAPILLOSE
truncated peristome
Dicranum viride
Drummondia prorepens
Dicranum montanum
Untoothed Leaves
CELLS UNI-PAPILLOSE
CELLS PLURI-PAPILLOSE
shortly exserted capsules
U. crispa capsule
U. coarctata capsule
capsules barely emergent
multicellular globose gemmae on costa
Syntrichia papillosa
Ulota crispa
Ulota coarctata
Orthotrichum sp.
(see pg. 193)

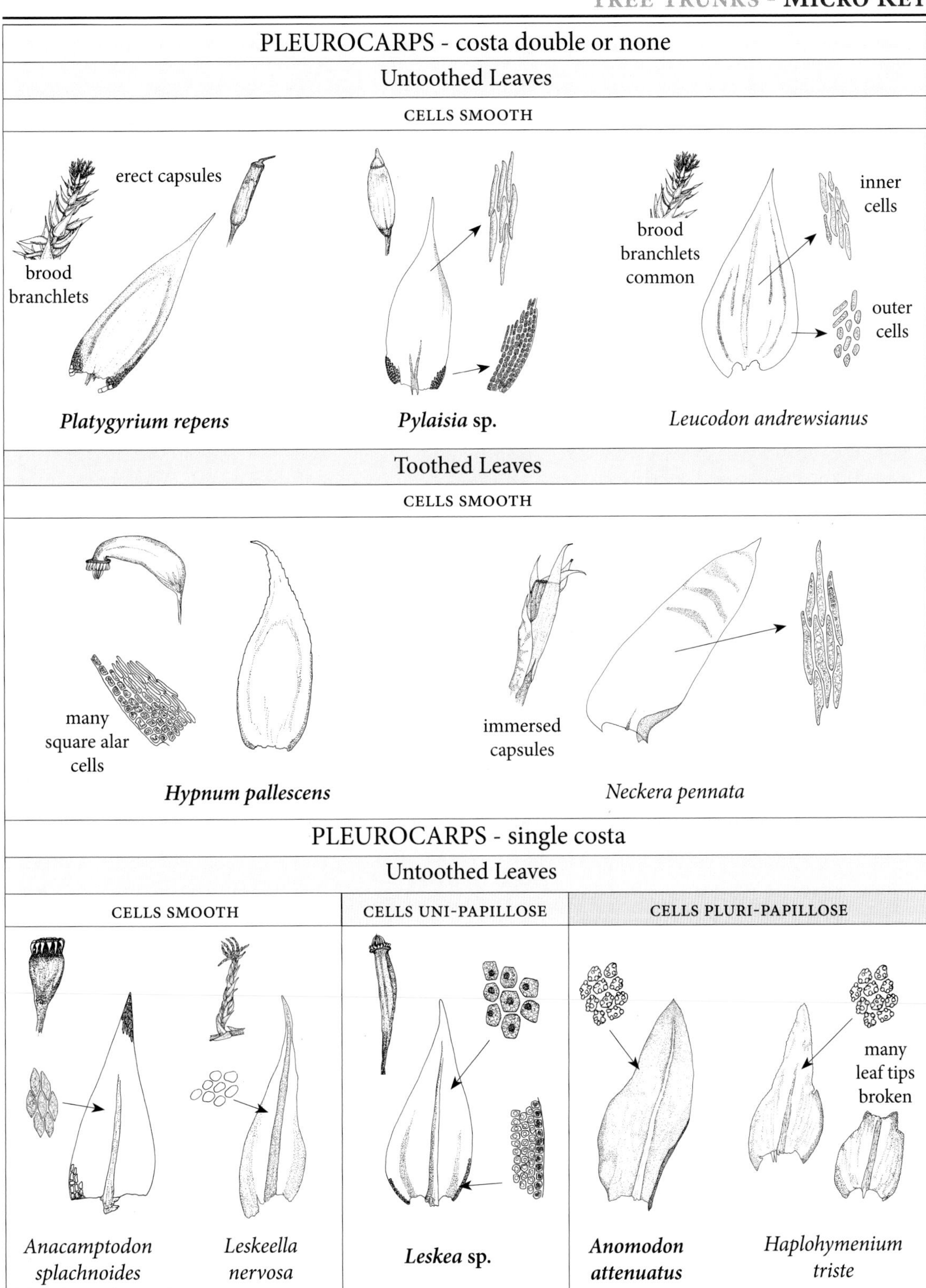
PLEUROCARPS - costa double or none
Untoothed Leaves
CELLS SMOOTH
erect capsules
brood
branchlets
Platygyrium repens
Pylaisia sp.
brood
branchlets
common
inner
cells
outer
cells
Leucodon andrewsianus
Toothed Leaves
CELLS SMOOTH
many
square alar
cells
Hypnum pallescens
immersed
capsules
Neckera pennata
PLEUROCARPS - single costa
Untoothed Leaves
CELLS SMOOTH
CELLS UNI-PAPILLOSE
CELLS PLURI-PAPILLOSE
Anacamptodon
splachnoides
Leskeella
nervosa
Leskea sp.
many
leaf tips
broken
Anomodon
attenuatus
Haplohymenium
triste

Dicranum montanum ***Ulota crispa*** *Ulota coarctata*

Orthotrichum sordidum

Orthotrichum stellatum

Orthotrichum obtusifolium

Dicranum viride

Drummondia prorepens

Leskea gracilescens - dry & wet

Anomodon attenuatus - dry & wet

Platygyrium repens

Hypnum pallescens

Anacamptodon splachnoides

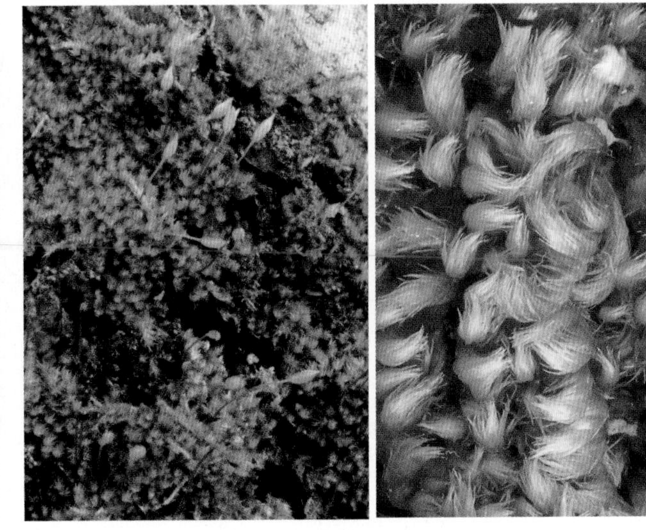

Pylaisia selwynii

Neckera pennata

Leucodon andrewsianus

Liverworts

Porella platyphylla

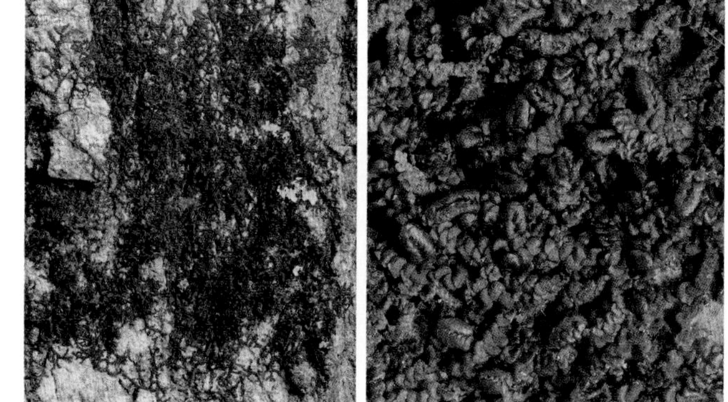

Frullania eboracensis

Ptilidium pulcherrimum

Radula complanata

TREE BASES

TREE BASES

As your eyes travel down the trunk of the tree towards its base, you may notice that the bryophyte abundance increases. Near the forest floor, there is little wind and more moisture available from the soil. Bryophytes care more about moisture levels than light levels for optimal growth, and mesic forests with high moisture content will have their bases entirely enveloped in a moss sock while drier woods might have only sparse coverage.

Mosses are small, and not many are seen on the forest floor where they would be covered by leaves. So they stick to areas that keep them above the litter - tree bases, logs, stumps and rocks help keep the mosses from being covered.

In moister habitats, such as a boreal forest or in a forested swamp where the moisture levels remain high, the distinction between tree base and ground becomes much less distinct, with many of the ground mosses climbing up the tree bases. This can also happen when a tree grows over a ledge or boulder, as seen below.

Did the *Anomodon attenuatus* start on the tree base and move to the rock, or did it start on the rock and creep up onto the tree base? Whichever way it went, it seems to be happy on either substrate.

Like the tree trunk species, the type of tree influences not only the species found, but the diversity and coverage as well. The ash in this second growth woods had a fairly high coverage of *Callicladium haldanianum* and *Platygyrium repens*, while the pine was only sparsely covered with *Dicranum montanum* and *Hypnum pallescens*.

Most Common Bryophytes of Pine/Hemlock Bases

Font size indicates relative abundance on tree bases

Dicranum montanum
Hypnum pallescens

Plagiothecium laetum
Ptilidium pulcherrimum

Most Common Bryophytes of Second Growth Hardwood Bases

Dicranum montanum
Hypnum pallescens
Platygyrium repens
Ptilidium pulcherrimum

Brachythecium campestre
Hypnum imponens
Lophocolea heterophylla
Thuidium delicatulum

Brachythecium reflexum

Callicladium haldanianum

Amblystegium varium
Dicranum viride
Syzygiella autumnalis
Plagiothecium laetum
Radula complanata
Ulota crispa

Most Common Bryophytes of Old Growth Hardwood Bases

Brachythecium reflexum
Porella platyphylla

Anomodon attenuatus
Hypnum pallescens
Plagiomnium cuspidatum
Ulota crispa

Brachythecium laetum
Dicranum viride

Anomodon rugelii
Platygyrium repens
Thuidium delicatulum
Rauiella scita

Leskeella nervosa
Radula complanata

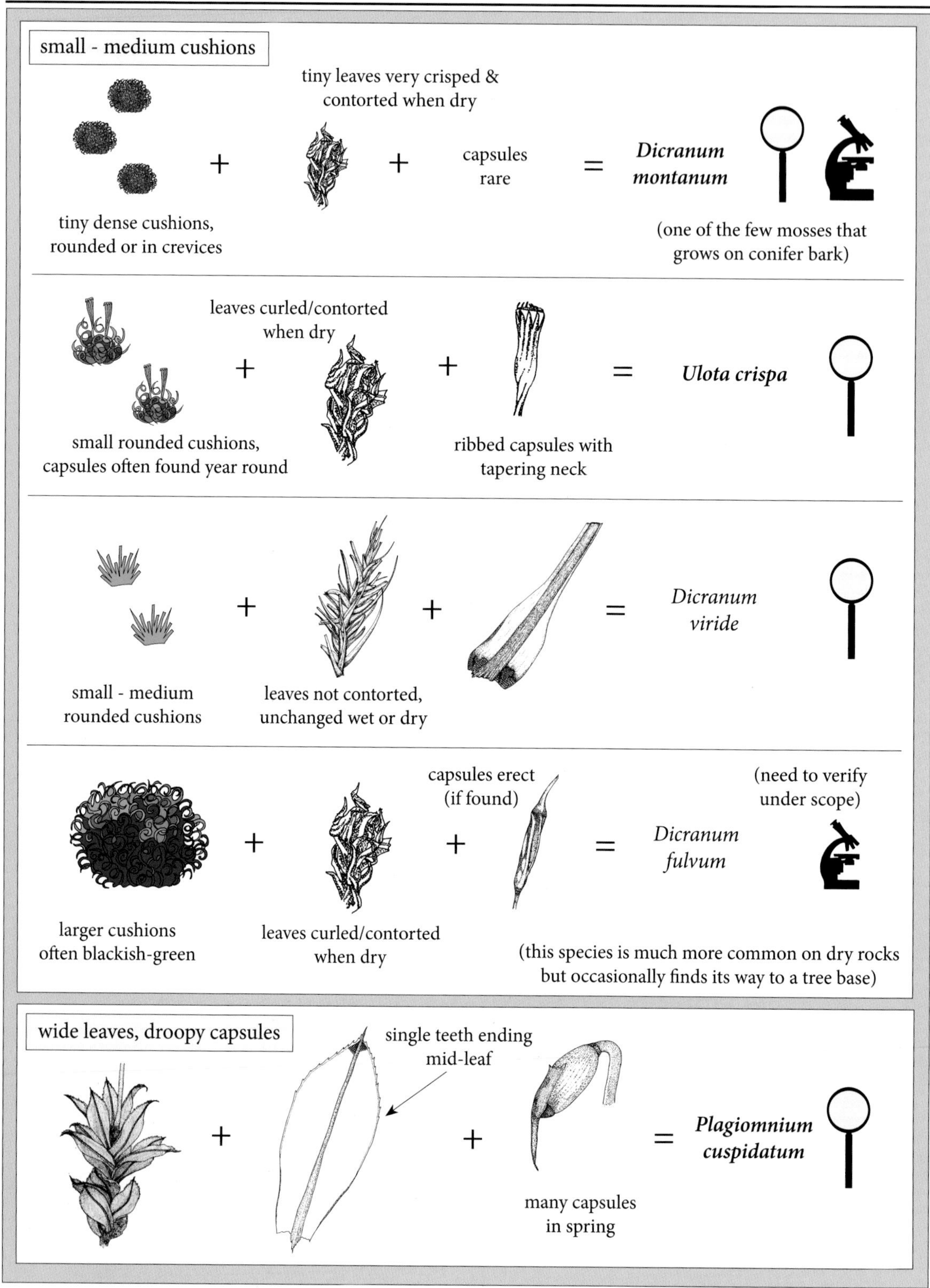
small - medium cushions
tiny dense cushions, rounded or in crevices
tiny leaves very crisped & contorted when dry
capsules rare
Dicranum montanum
(one of the few mosses that grows on conifer bark)
small rounded cushions, capsules often found year round
leaves curled/contorted when dry
ribbed capsules with tapering neck
Ulota crispa
small - medium rounded cushions
leaves not contorted, unchanged wet or dry
Dicranum viride
larger cushions often blackish-green
leaves curled/contorted when dry
capsules erect (if found)
Dicranum fulvum
(need to verify under scope)
(this species is much more common on dry rocks but occasionally finds its way to a tree base)
wide leaves, droopy capsules
single teeth ending mid-leaf
many capsules in spring
Plagiomnium cuspidatum

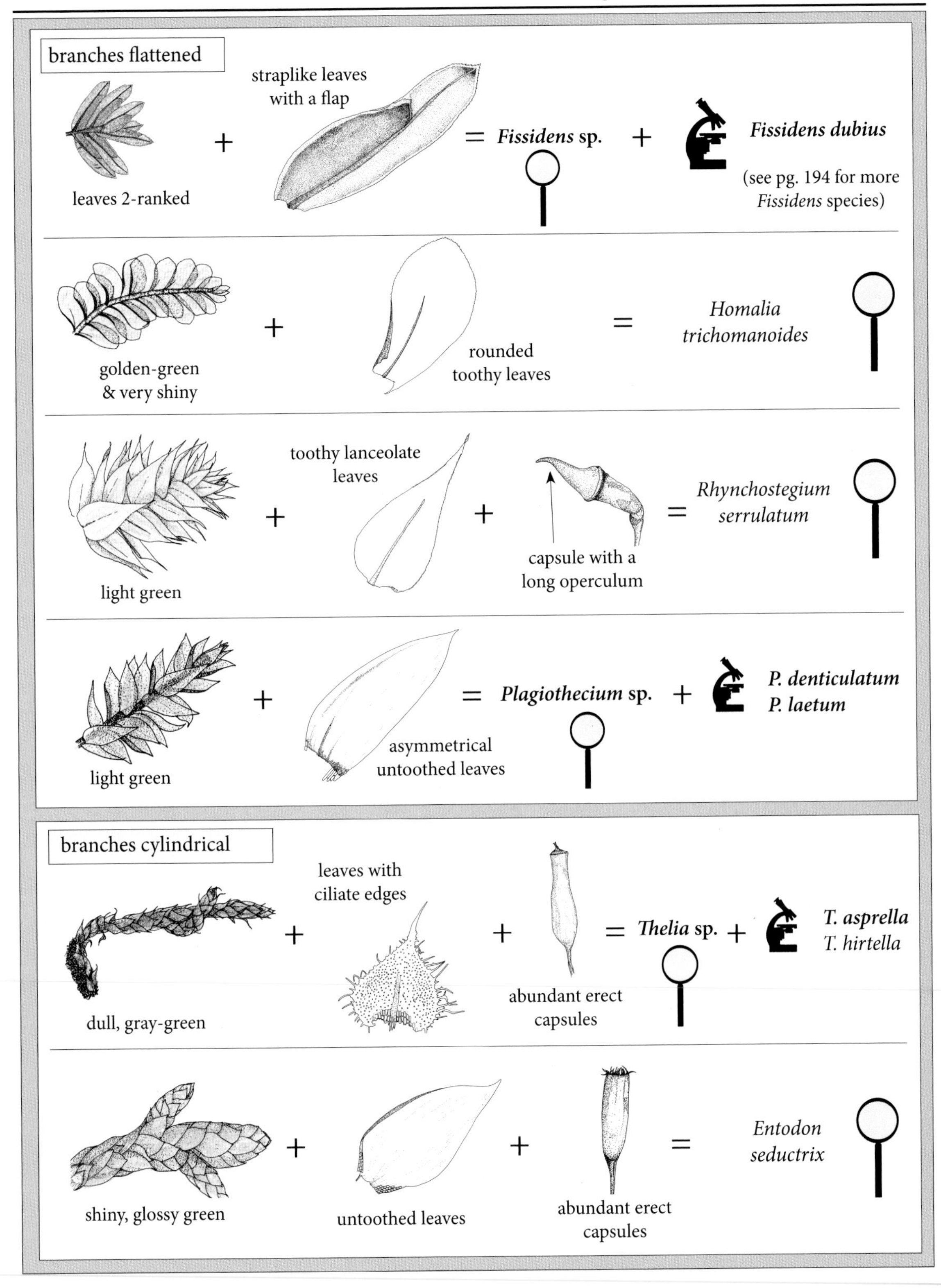
branches flattened
leaves 2-ranked
straplike leaves with a flap
= Fissidens sp.
Fissidens dubius
(see pg. 194 for more Fissidens species)
golden-green & very shiny
rounded toothy leaves
= Homalia trichomanoides
light green
toothy lanceolate leaves
capsule with a long operculum
= Rhynchostegium serrulatum
light green
asymmetrical untoothed leaves
= Plagiothecium sp.
P. denticulatum
P. laetum
branches cylindrical
dull, gray-green
leaves with ciliate edges
abundant erect capsules
= Thelia sp.
T. asprella
T. hirtella
shiny, glossy green
untoothed leaves
abundant erect capsules
= Entodon seductrix

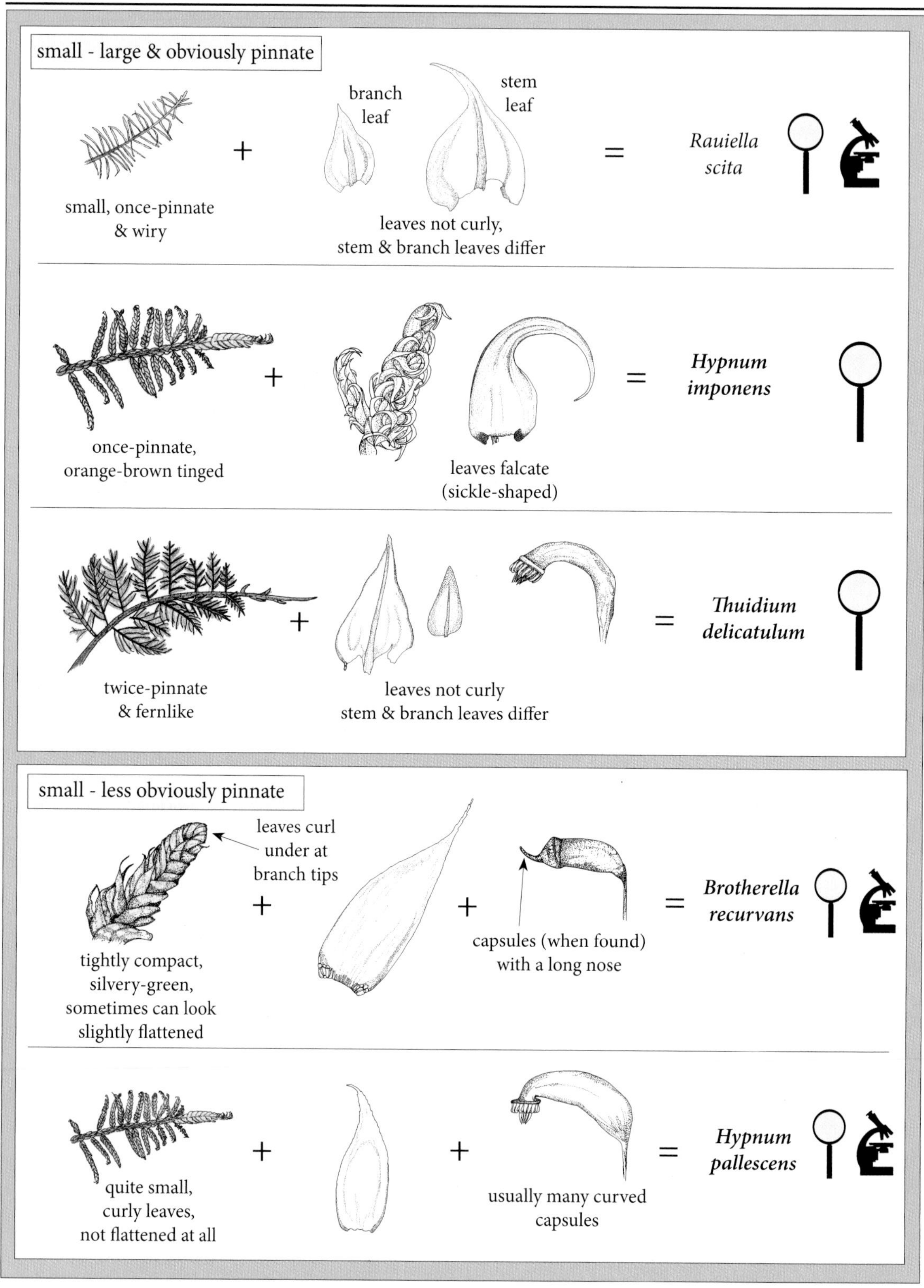
small - large & obviously pinnate
small, once-pinnate & wiry
+
branch leaf
stem leaf
leaves not curly, stem & branch leaves differ
=
Rauiella scita
once-pinnate, orange-brown tinged
+
leaves falcate (sickle-shaped)
=
Hypnum imponens
twice-pinnate & fernlike
+
leaves not curly stem & branch leaves differ
=
Thuidium delicatulum
small - less obviously pinnate
leaves curl under at branch tips
tightly compact, silvery-green, sometimes can look slightly flattened
+
+
capsules (when found) with a long nose
=
Brotherella recurvans
quite small, curly leaves, not flattened at all
+
+
usually many curved capsules
=
Hypnum pallescens

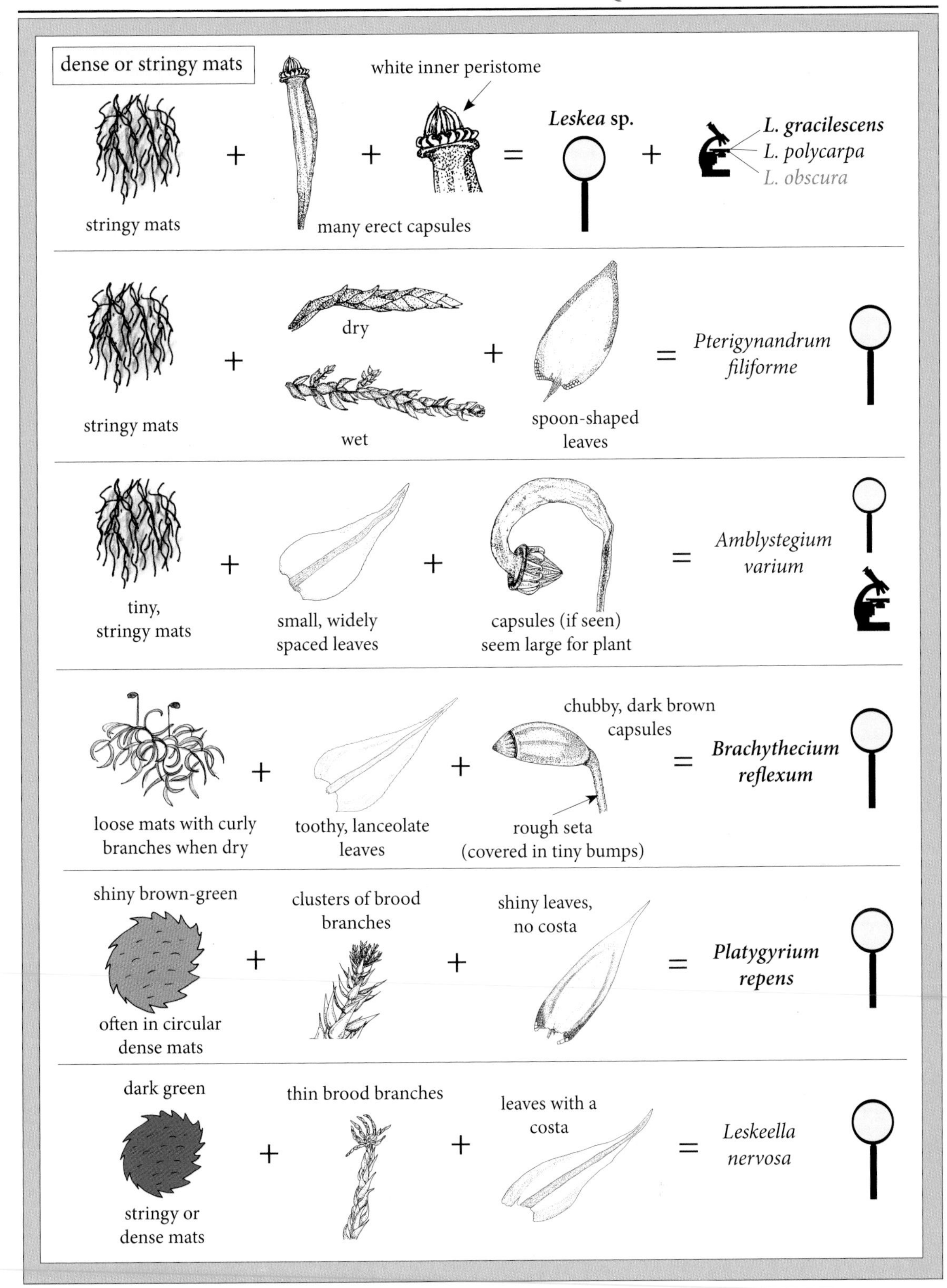
dense or stringy mats
stringy mats
many erect capsules
white inner peristome
Leskea sp.
L. gracilescens
L. polycarpa
L. obscura
stringy mats
dry
wet
spoon-shaped leaves
Pterigynandrum filiforme
tiny, stringy mats
small, widely spaced leaves
capsules (if seen) seem large for plant
Amblystegium varium
loose mats with curly branches when dry
toothy, lanceolate leaves
chubby, dark brown capsules
rough seta (covered in tiny bumps)
Brachythecium reflexum
shiny brown-green
often in circular dense mats
clusters of brood branches
shiny leaves, no costa
Platygyrium repens
dark green
stringy or dense mats
thin brood branches
leaves with a costa
Leskeella nervosa

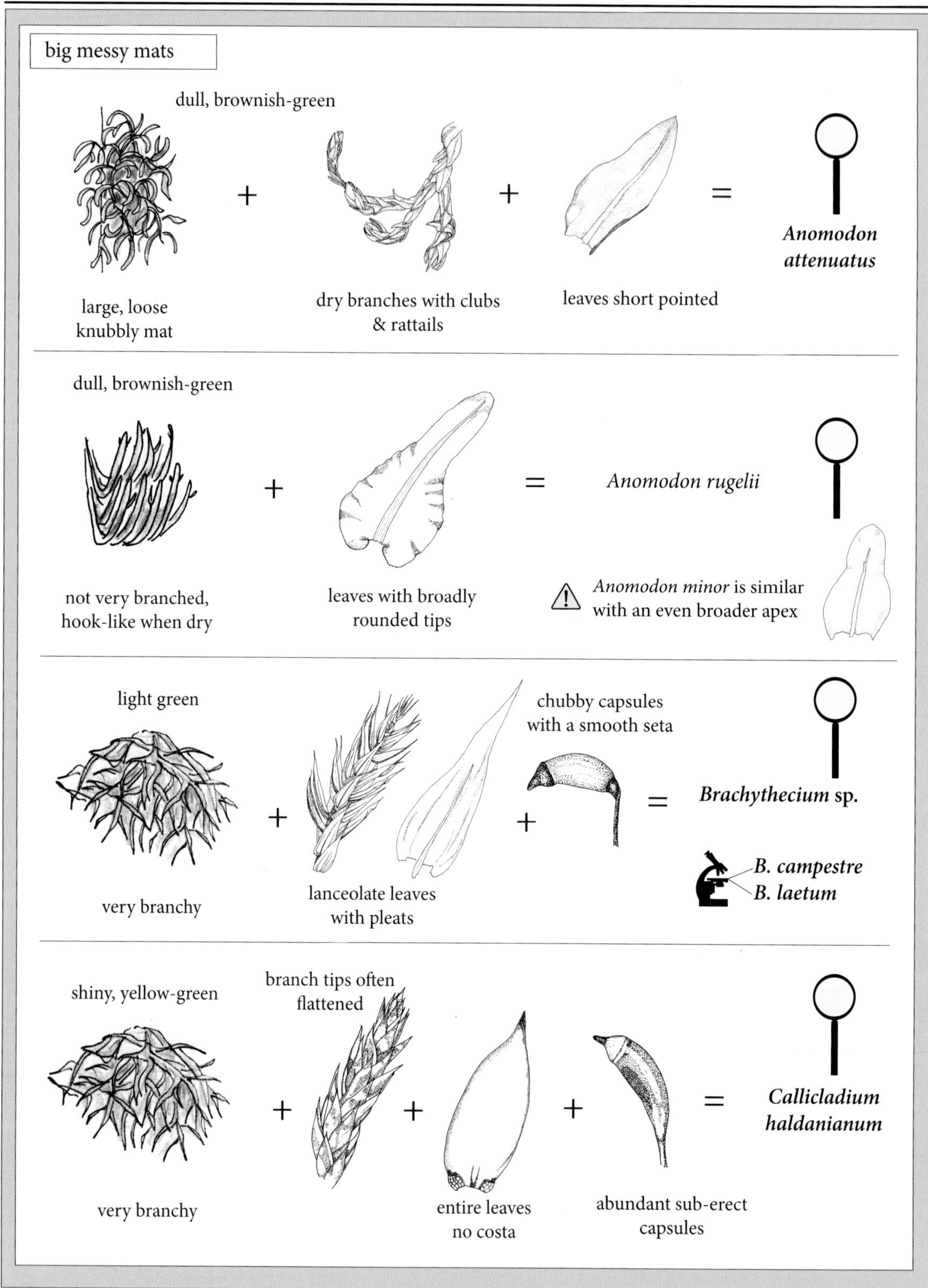
big messy mats
dull, brownish-green
large, loose knubbly mat
dry branches with clubs & rattails
leaves short pointed
Anomodon attenuatus
dull, brownish-green
not very branched, hook-like when dry
leaves with broadly rounded tips
Anomodon rugelii
Anomodon minor is similar with an even broader apex
light green
very branchy
lanceolate leaves with pleats
chubby capsules with a smooth seta
Brachythecium sp.
B. campestre
B. laetum
shiny, yellow-green
very branchy
branch tips often flattened
entire leaves no costa
abundant sub-erect capsules
Callicladium haldanianum

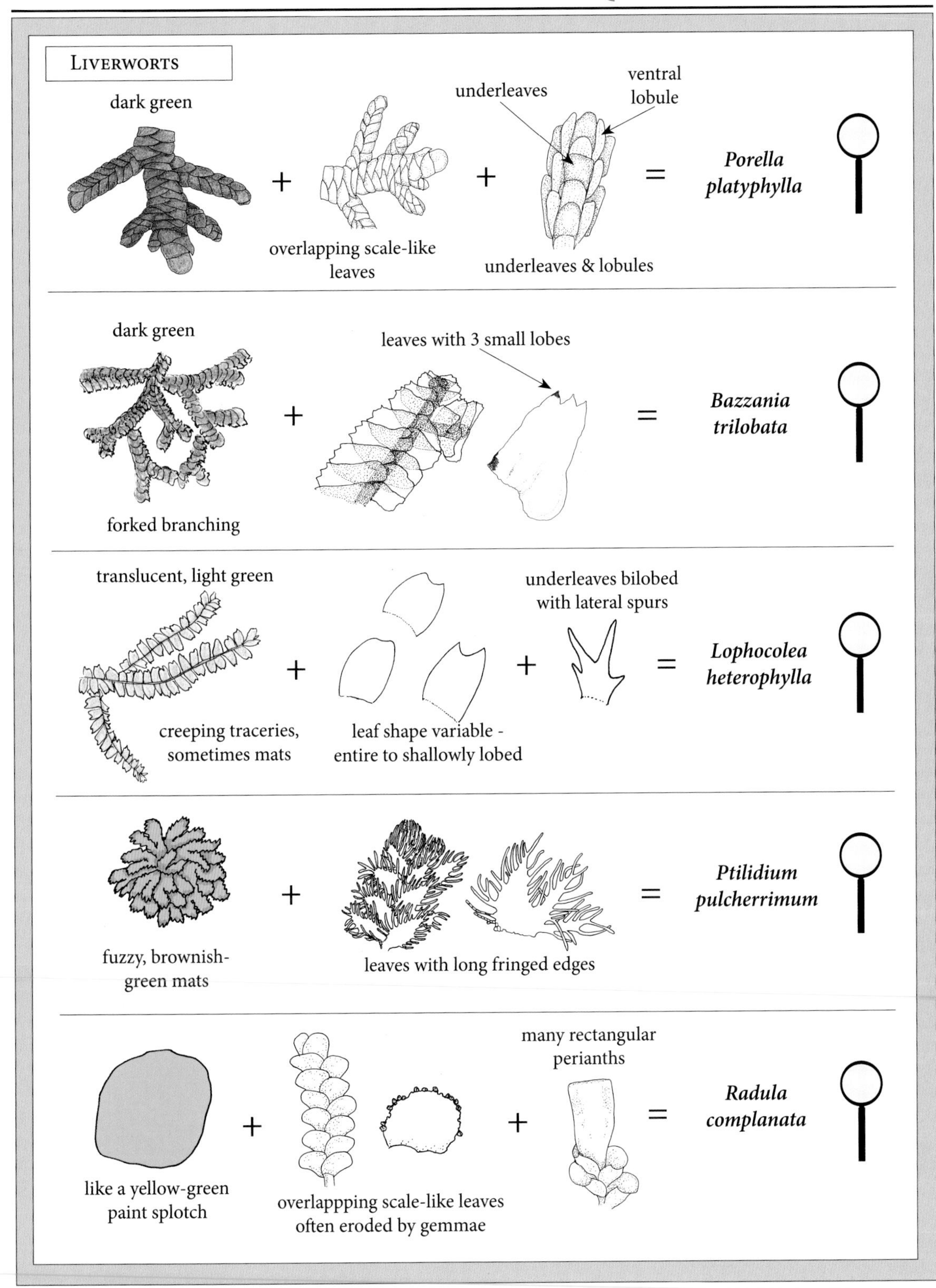
LIVERWORTS
dark green
underleaves
ventral lobule
overlapping scale-like leaves
underleaves & lobules
Porella platyphylla
dark green
leaves with 3 small lobes
forked branching
Bazzania trilobata
translucent, light green
underleaves bilobed with lateral spurs
creeping traceries, sometimes mats
leaf shape variable - entire to shallowly lobed
Lophocolea heterophylla
fuzzy, brownish-green mats
leaves with long fringed edges
Ptilidium pulcherrimum
many rectangular perianths
like a yellow-green paint splotch
overlappping scale-like leaves often eroded by gemmae
Radula complanata

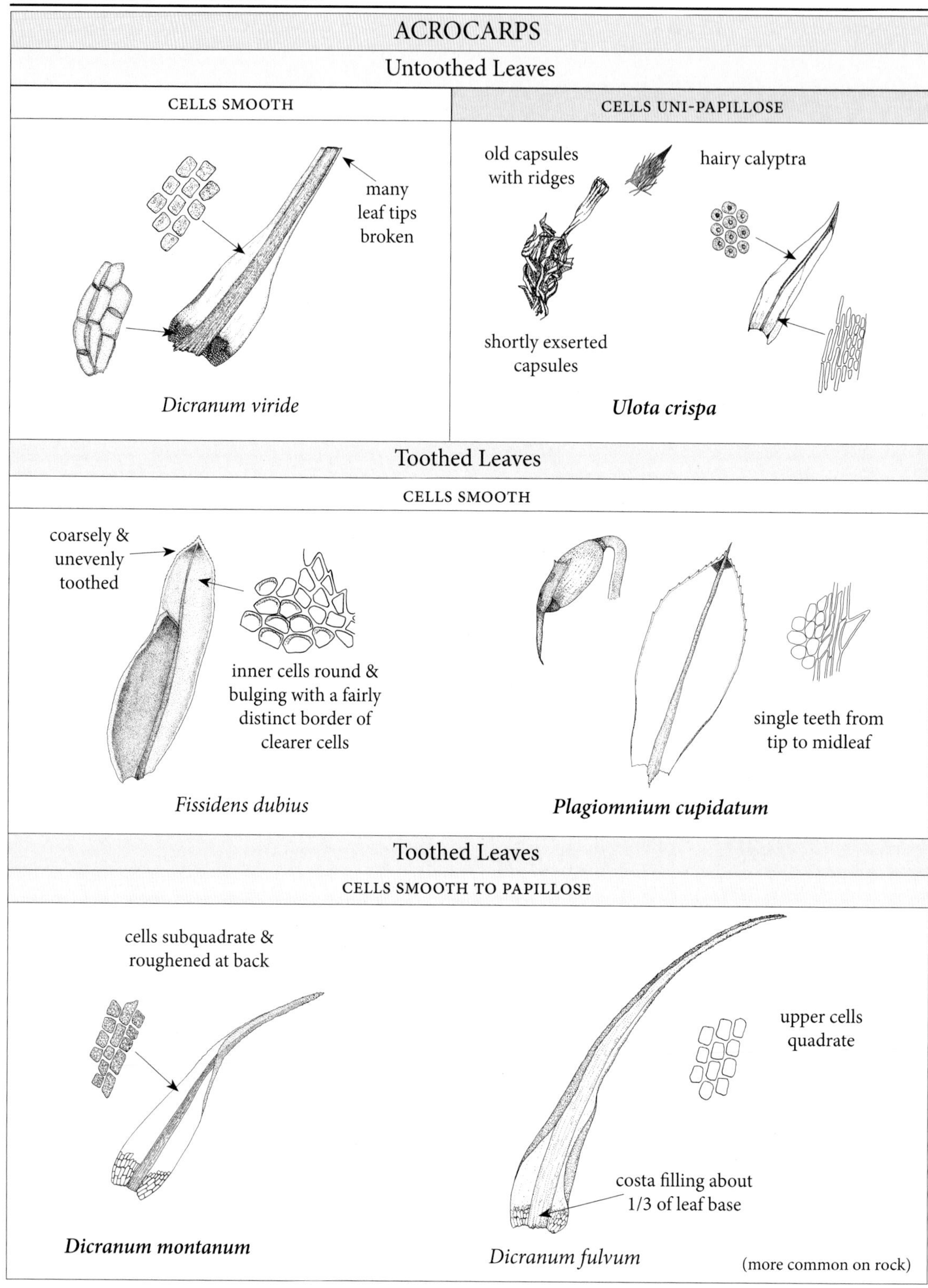
ACROCARPS
Untoothed Leaves
CELLS SMOOTH
CELLS UNI-PAPILLOSE
many leaf tips broken
Dicranum viride
old capsules with ridges
hairy calyptra
shortly exserted capsules
Ulota crispa
Toothed Leaves
CELLS SMOOTH
coarsely & unevenly toothed
inner cells round & bulging with a fairly distinct border of clearer cells
Fissidens dubius
single teeth from tip to midleaf
Plagiomnium cupidatum
Toothed Leaves
CELLS SMOOTH TO PAPILLOSE
cells subquadrate & roughened at back
Dicranum montanum
upper cells quadrate
costa filling about 1/3 of leaf base
Dicranum fulvum
(more common on rock)

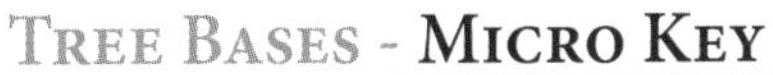

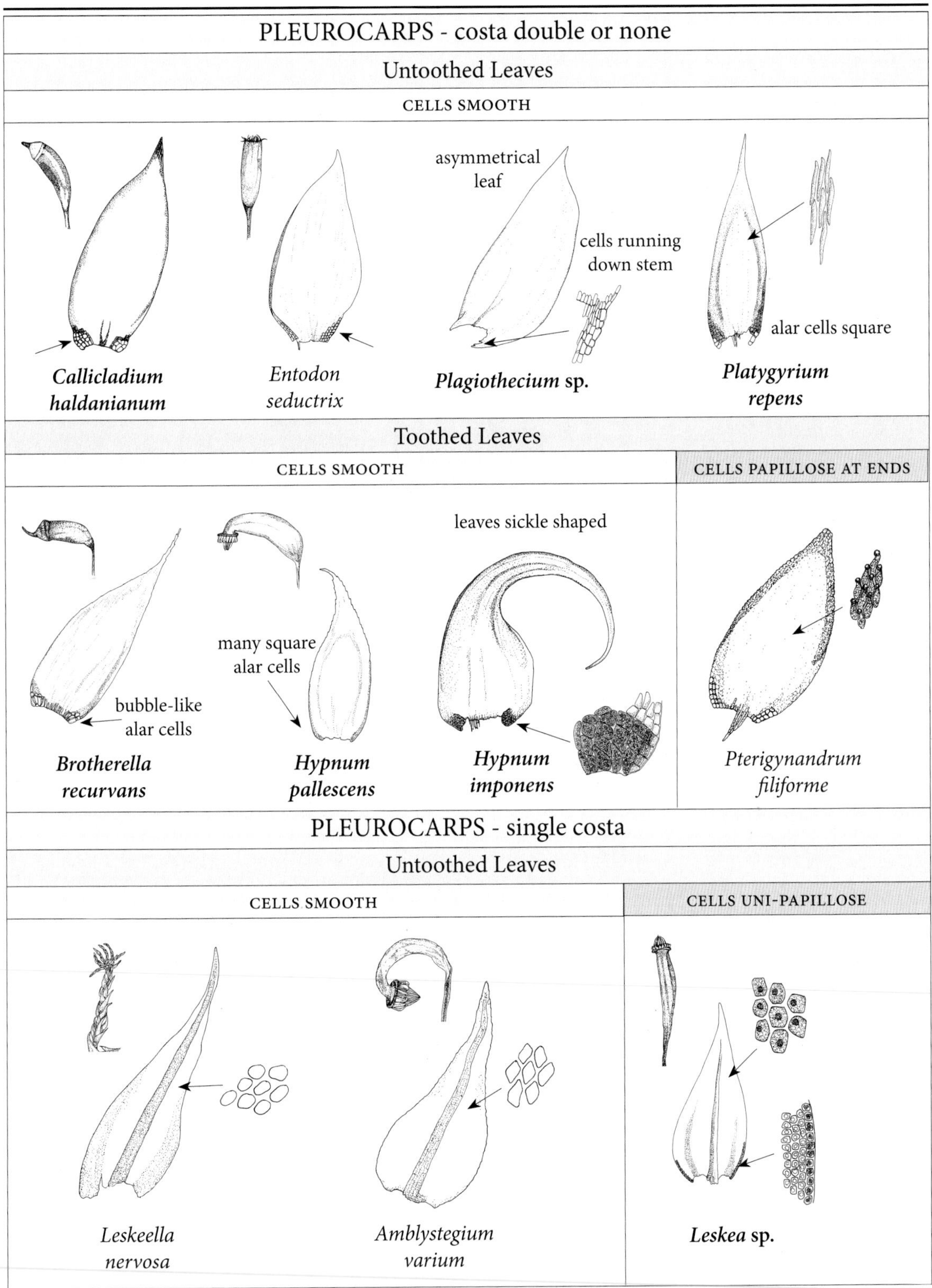
PLEUROCARPS - costa double or none
Untoothed Leaves
CELLS SMOOTH
asymmetrical leaf
cells running down stem
alar cells square
Callicladium haldanianum
Entodon seductrix
Plagiothecium sp.
Platygyrium repens
Toothed Leaves
CELLS SMOOTH
CELLS PAPILLOSE AT ENDS
leaves sickle shaped
many square alar cells
bubble-like alar cells
Brotherella recurvans
Hypnum pallescens
Hypnum imponens
Pterigynandrum filiforme
PLEUROCARPS - single costa
Untoothed Leaves
CELLS SMOOTH
CELLS UNI-PAPILLOSE
Leskeella nervosa
Amblystegium varium
Leskea sp.

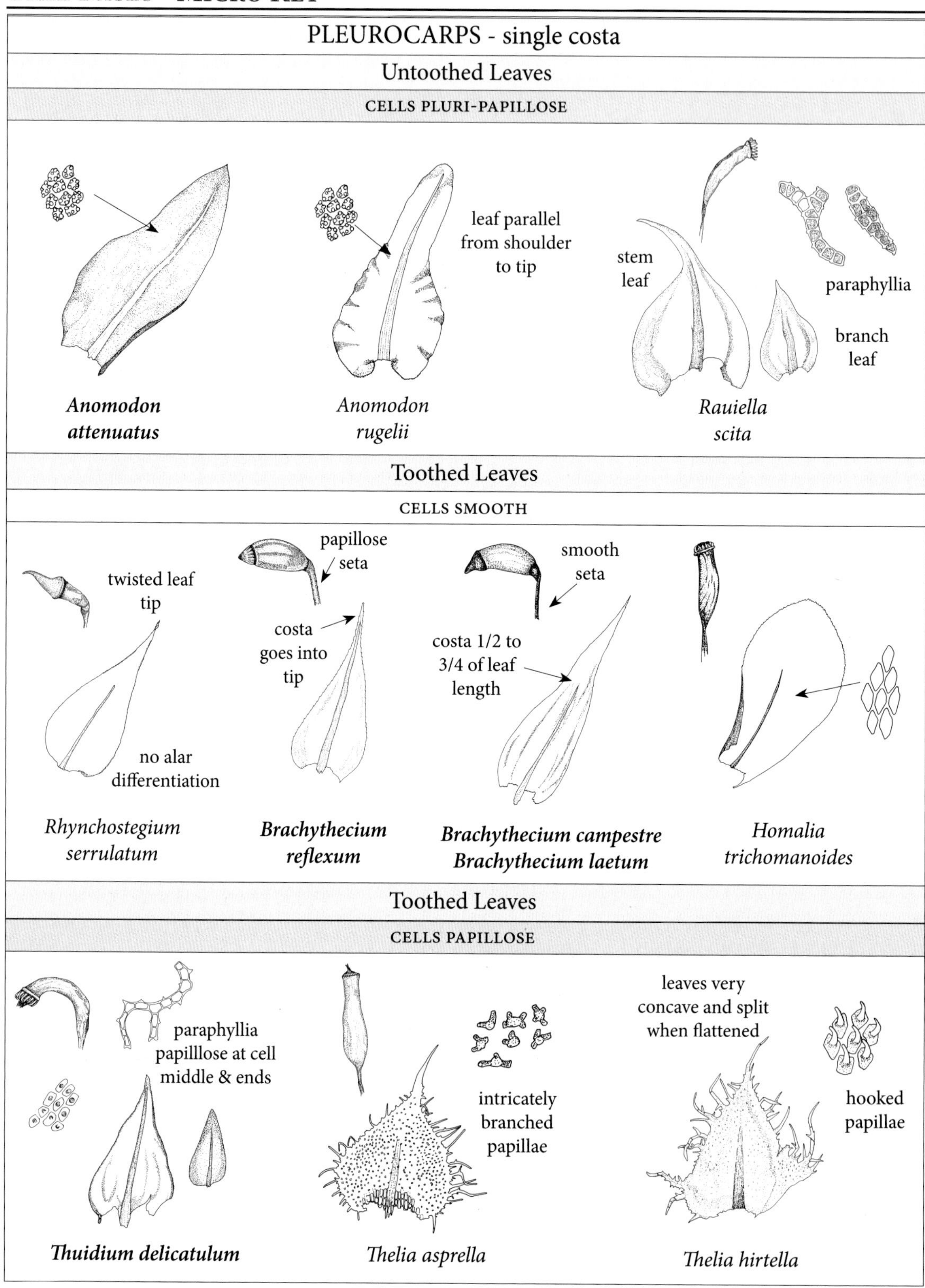
PLEUROCARPS - single costa
Untoothed Leaves
CELLS PLURI-PAPILLOSE
leaf parallel from shoulder to tip
stem leaf
paraphyllia
branch leaf
Anomodon attenuatus
Anomodon rugelii
Rauiella scita
Toothed Leaves
CELLS SMOOTH
twisted leaf tip
papillose seta
smooth seta
costa goes into tip
costa 1/2 to 3/4 of leaf length
no alar differentiation
Rhynchostegium serrulatum
Brachythecium reflexum
Brachythecium campestre
Brachythecium laetum
Homalia trichomanoides
Toothed Leaves
CELLS PAPILLOSE
paraphyllia papilllose at cell middle & ends
leaves very concave and split when flattened
intricately branched papillae
hooked papillae
Thuidium delicatulum
Thelia asprella
Thelia hirtella

Dicranum montanum

Dicranum viride

Dicranum fulvum

Plagiomnium cuspidatum

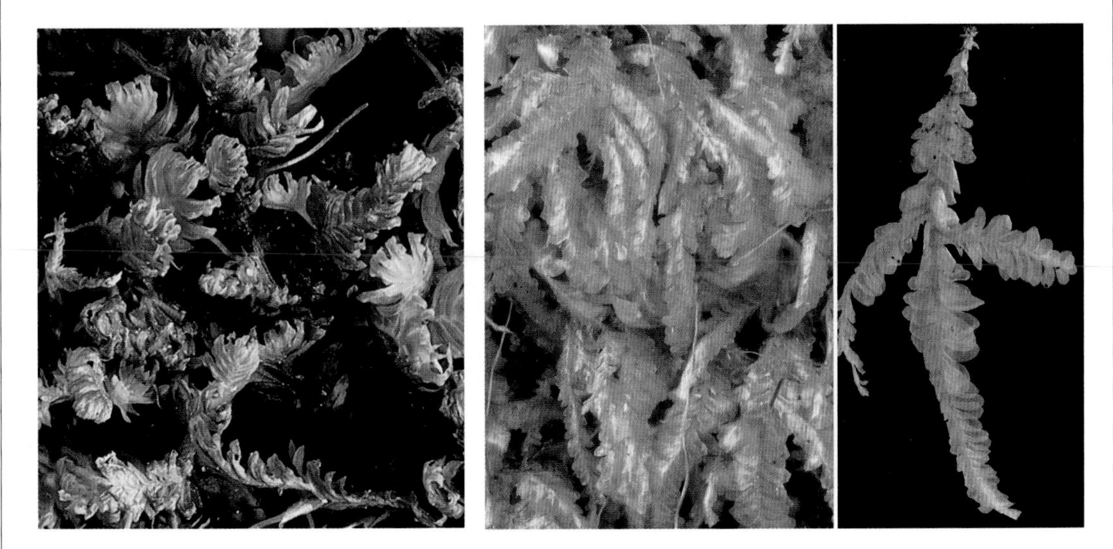

Fissidens dubius

Homalia trichomanoides

Plagiothecium denticulatum

Rhynchostegium serrulatum

Thelia asprella

Entodon seductrix

Pterigynandrum filiforme

Hypnum imponens

Hypnum pallescens

Rauiella scita

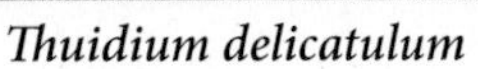

Thuidium delicatulum

Brotherella recurvans

Brachythecium reflexum

Brachythecium campestre/laetum

Callicladium haldanianum

Liverworts

Porella platyphylla

Bazzania trilobata

Ptilidium pulcherrimum

Radula complanata

ROTTEN LOGS
(SAW)
&
STUMPS
(SAW)

When a tree falls in the woods, whether by lightning, beaver, woodchopper or old age, what happens to the bryophytes? If the woods are dry, not much. After a few years, the mosses that were found on the trunk eventually die. If, however, the tree falls onto moist ground, decay begins, and the tree starts to rot. At first, the log still retains most of its bark, and the mosses found on newly fallen trees are the mosses you would find on tree trunks, like the small cushions of *Ulota crispa* seen on this birch log.

(SAW)

As the log deteriorates further, the bark falls off, the wood softens and different mosses and liverworts will now be found on it. It appears at this point that whichever moss gets started the quickest is the one that will cover the log. Some logs are entirely covered in one species - *Callicladium haldanianum*, *Hypnum imponens*, and *Thuidium delicatulum* are common species that can cover entire logs, as well as the liverwort *Nowellia curvifolia*. Some logs, however, have quite a diversity of bryophytes and can have 10 different species. Liverworts appreciate the sodden, decorticated logs for the ample and consistent moisture levels found. Because logs are a transient substrate, going from tree trunk and ending up as humus, many of the bryophytes found on them are also found on other substrates, especially tree bases and humic soil. There are only a few species that are not found on something else. *Buxbaumia minakatae* is a very rare moss of logs (perhaps overlooked?). *Oncophorus wahlenbergii*, a member of the *Dicranum* family, likes logs in particularly wet areas, as does *Plagiothecium latebricola*. Look for these two species on logs in swampy places. *Dicranum flagellare* is a common species of drier areas, and while it can be found on other substrates, it is especially fond of logs. Several liverworts seem to prefer logs as well, including the ubiquitous *Nowellia curvifolia* and *Lophocolea heterophylla*. Less frequently encountered are *Riccardia latifrons*, *R. palmata* and *Odontoschisma denudatum*, lovers of sodden logs.

These two photos illustrate the further decay of a log. The log on the left is now fully decorticated (all the bark is gone), and it has a complete covering of *Thuidium delicatulum*. The photo on the right shows the eventual decay of a log into soil.

(SAW)

(SAW)

A stump left from loggers also decays, and the tree stump specialist in our area is *Tetraphis pellucida*. If found on logs at all, it will be on the vertical sides, hardly ever on the top. *Tetraphis* reproduces both vegetatively and sexually. In colonizing a newly exposed area, it propagates quickly using gemmae splashed out of a cup. When conditions get more crowded, it switches over to sexual reproduction, and the plants are crowded with many upright capsules bearing four teeth and looking like the beak of a bird.

(SAW)

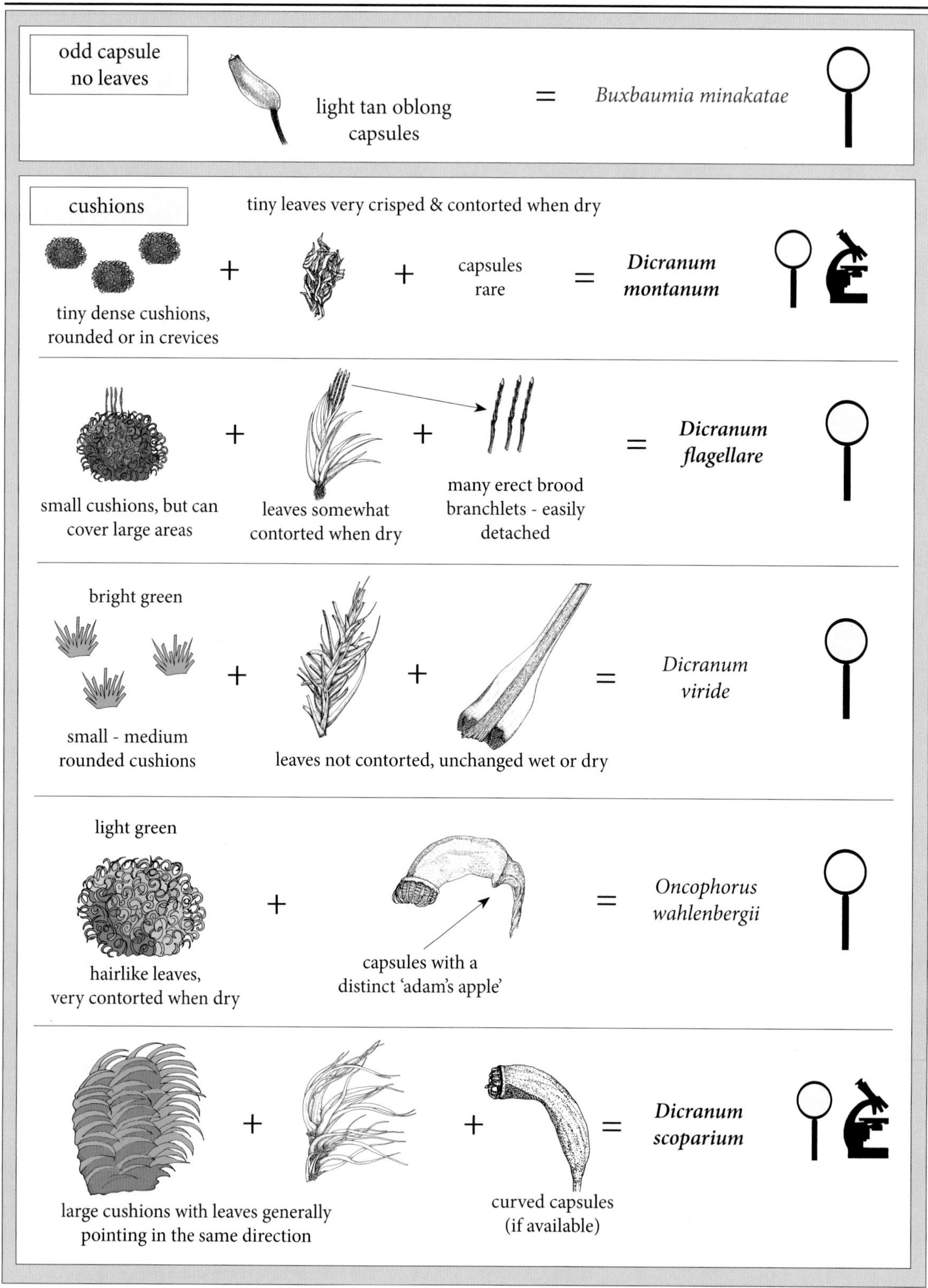
odd capsule
no leaves
light tan oblong
capsules
=
Buxbaumia minakatae
cushions
tiny leaves very crisped & contorted when dry
+
+
capsules
rare
=
Dicranum
montanum
tiny dense cushions,
rounded or in crevices
+
+
=
Dicranum
flagellare
small cushions, but can
cover large areas
leaves somewhat
contorted when dry
many erect brood
branchlets - easily
detached
bright green
+
+
=
Dicranum
viride
small - medium
rounded cushions
leaves not contorted, unchanged wet or dry
light green
+
=
Oncophorus
wahlenbergii
hairlike leaves,
very contorted when dry
capsules with a
distinct 'adam's apple'
+
+
=
Dicranum
scoparium
large cushions with leaves generally
pointing in the same direction
curved capsules
(if available)

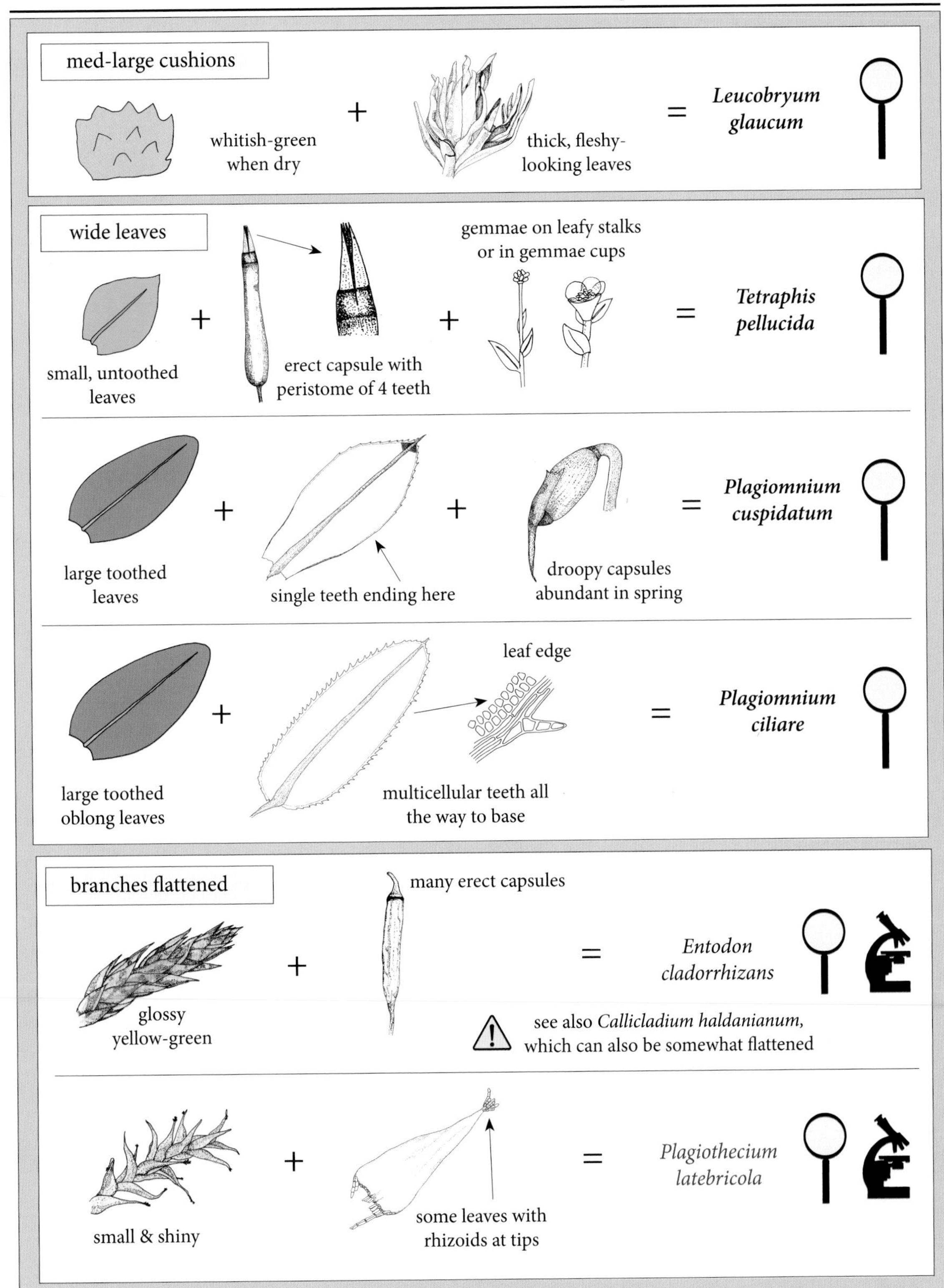
med-large cushions
whitish-green when dry
+
thick, fleshy-looking leaves
=
Leucobryum glaucum
wide leaves
small, untoothed leaves
+
erect capsule with peristome of 4 teeth
+
gemmae on leafy stalks or in gemmae cups
=
Tetraphis pellucida
large toothed leaves
+
single teeth ending here
+
droopy capsules abundant in spring
=
Plagiomnium cuspidatum
large toothed oblong leaves
+
multicellular teeth all the way to base
leaf edge
=
Plagiomnium ciliare
branches flattened
glossy yellow-green
+
many erect capsules
=
Entodon cladorrhizans
see also Callicladium haldanianum, which can also be somewhat flattened
small & shiny
+
some leaves with rhizoids at tips
=
Plagiothecium latebricola

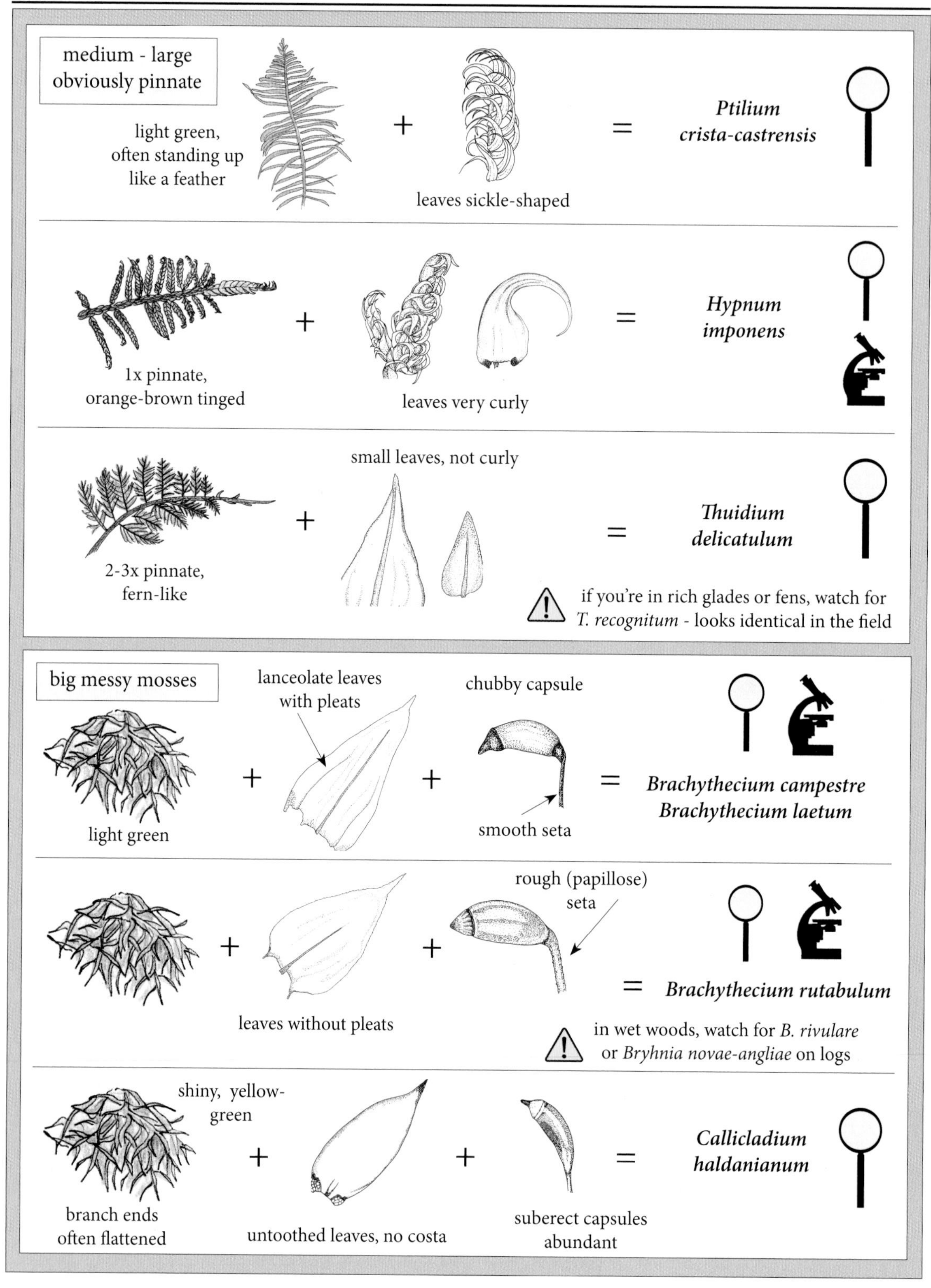
medium - large
obviously pinnate
light green,
often standing up
like a feather
+
leaves sickle-shaped
=
Ptilium crista-castrensis
1x pinnate,
orange-brown tinged
+
leaves very curly
=
Hypnum imponens
small leaves, not curly
2-3x pinnate,
fern-like
+
=
Thuidium delicatulum
if you're in rich glades or fens, watch for *T. recognitum* - looks identical in the field
big messy mosses
lanceolate leaves
with pleats
chubby capsule
light green
+
+
smooth seta
=
Brachythecium campestre
Brachythecium laetum
rough (papillose)
seta
+
+
leaves without pleats
=
Brachythecium rutabulum
in wet woods, watch for *B. rivulare* or *Bryhnia novae-angliae* on logs
shiny, yellow-
green
branch ends
often flattened
+
untoothed leaves, no costa
+
suberect capsules
abundant
=
Callicladium haldanianum

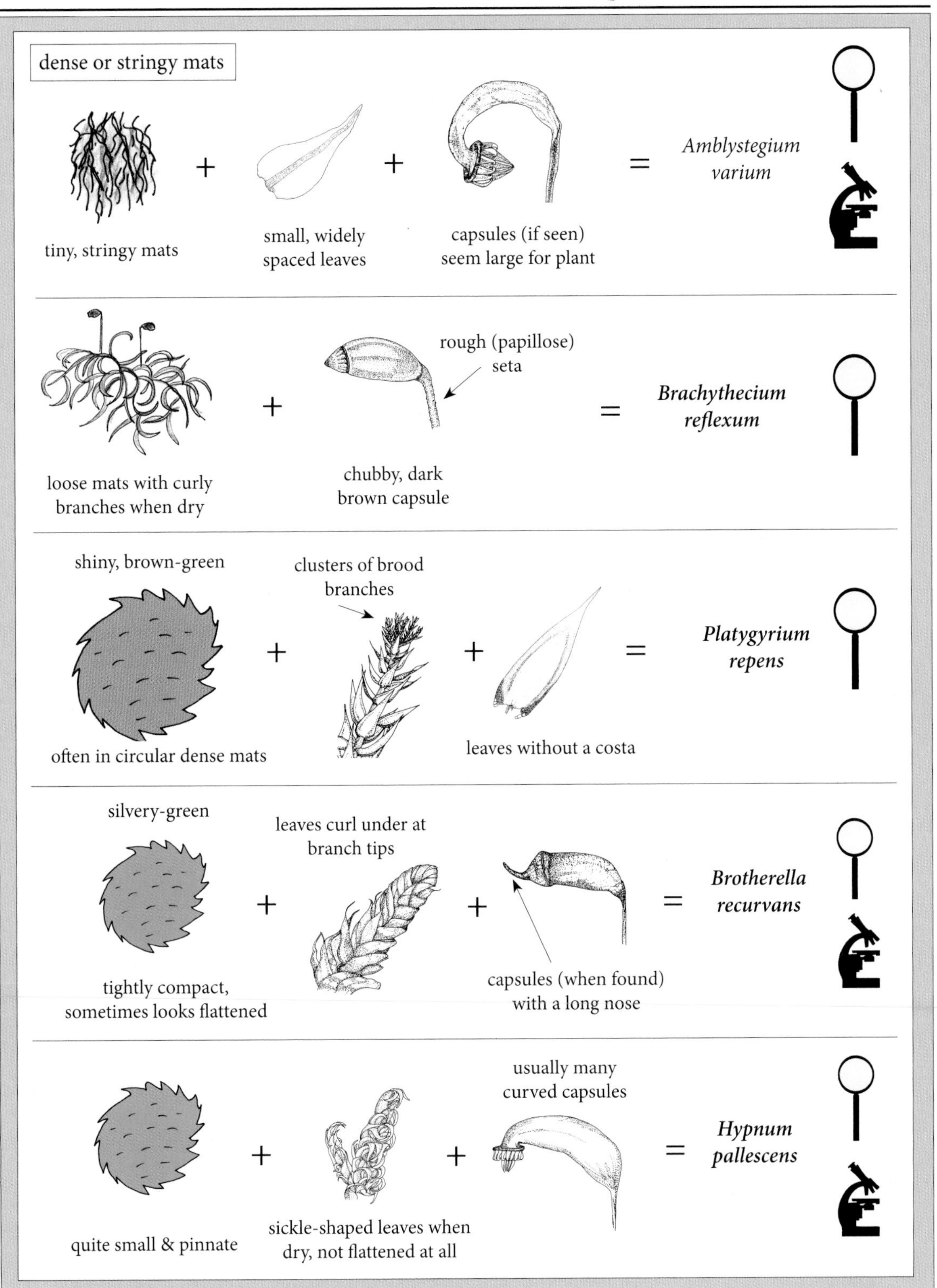
dense or stringy mats
tiny, stringy mats
+
small, widely spaced leaves
+
capsules (if seen) seem large for plant
=
Amblystegium varium
loose mats with curly branches when dry
+
chubby, dark brown capsule
rough (papillose) seta
=
Brachythecium reflexum
shiny, brown-green
often in circular dense mats
+
clusters of brood branches
+
leaves without a costa
=
Platygyrium repens
silvery-green
tightly compact, sometimes looks flattened
+
leaves curl under at branch tips
+
capsules (when found) with a long nose
=
Brotherella recurvans
quite small & pinnate
+
sickle-shaped leaves when dry, not flattened at all
+
usually many curved capsules
=
Hypnum pallescens

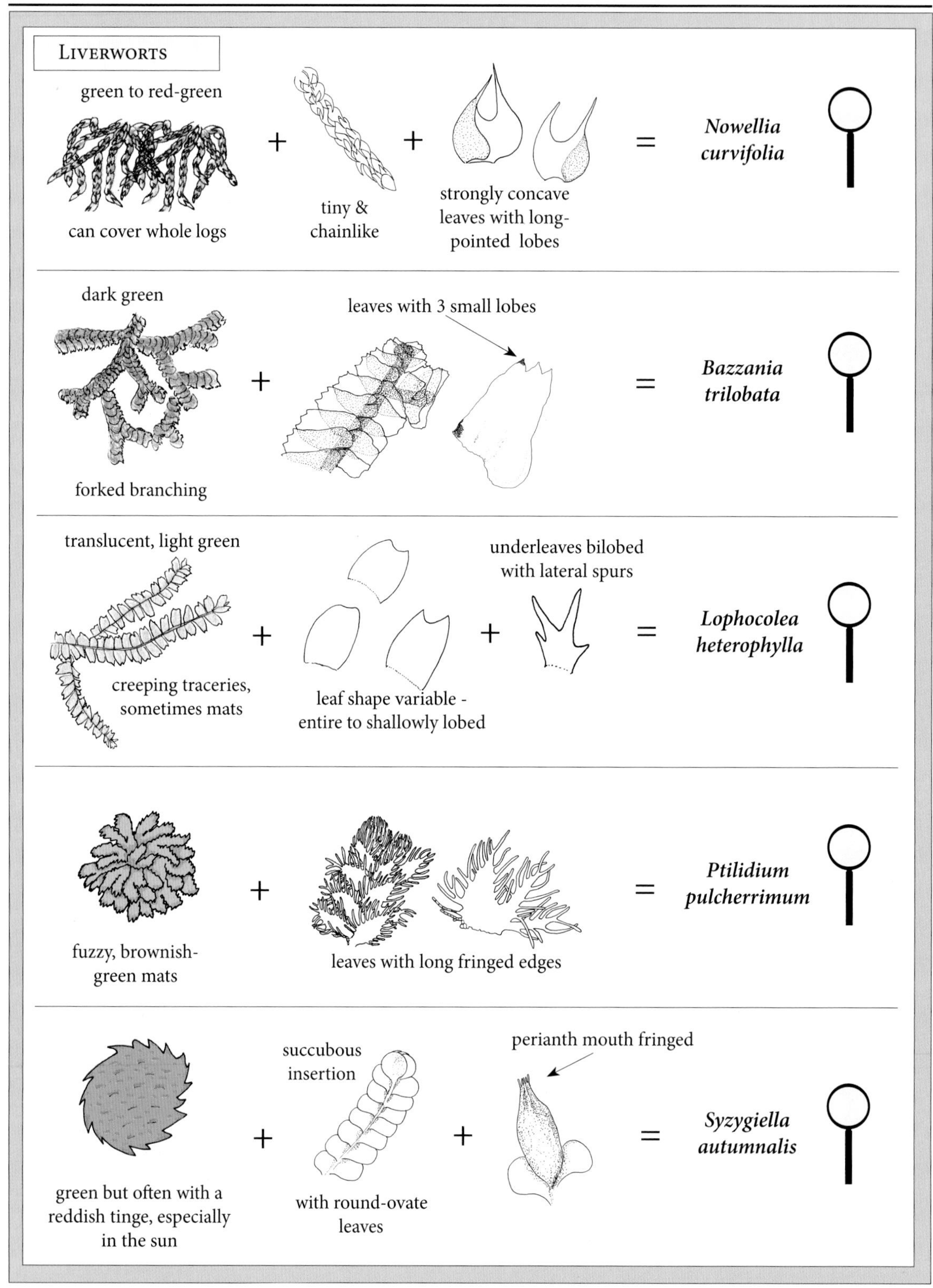

Liverworts
green to red-green
can cover whole logs
+
tiny & chainlike
+
strongly concave leaves with long-pointed lobes
=
Nowellia curvifolia
dark green
forked branching
+
leaves with 3 small lobes
=
Bazzania trilobata
translucent, light green
creeping traceries, sometimes mats
+
leaf shape variable - entire to shallowly lobed
+
underleaves bilobed with lateral spurs
=
Lophocolea heterophylla
fuzzy, brownish-green mats
+
leaves with long fringed edges
=
Ptilidium pulcherrimum
green but often with a reddish tinge, especially in the sun
+
succubous insertion
with round-ovate leaves
+
perianth mouth fringed
=
Syzygiella autumnalis

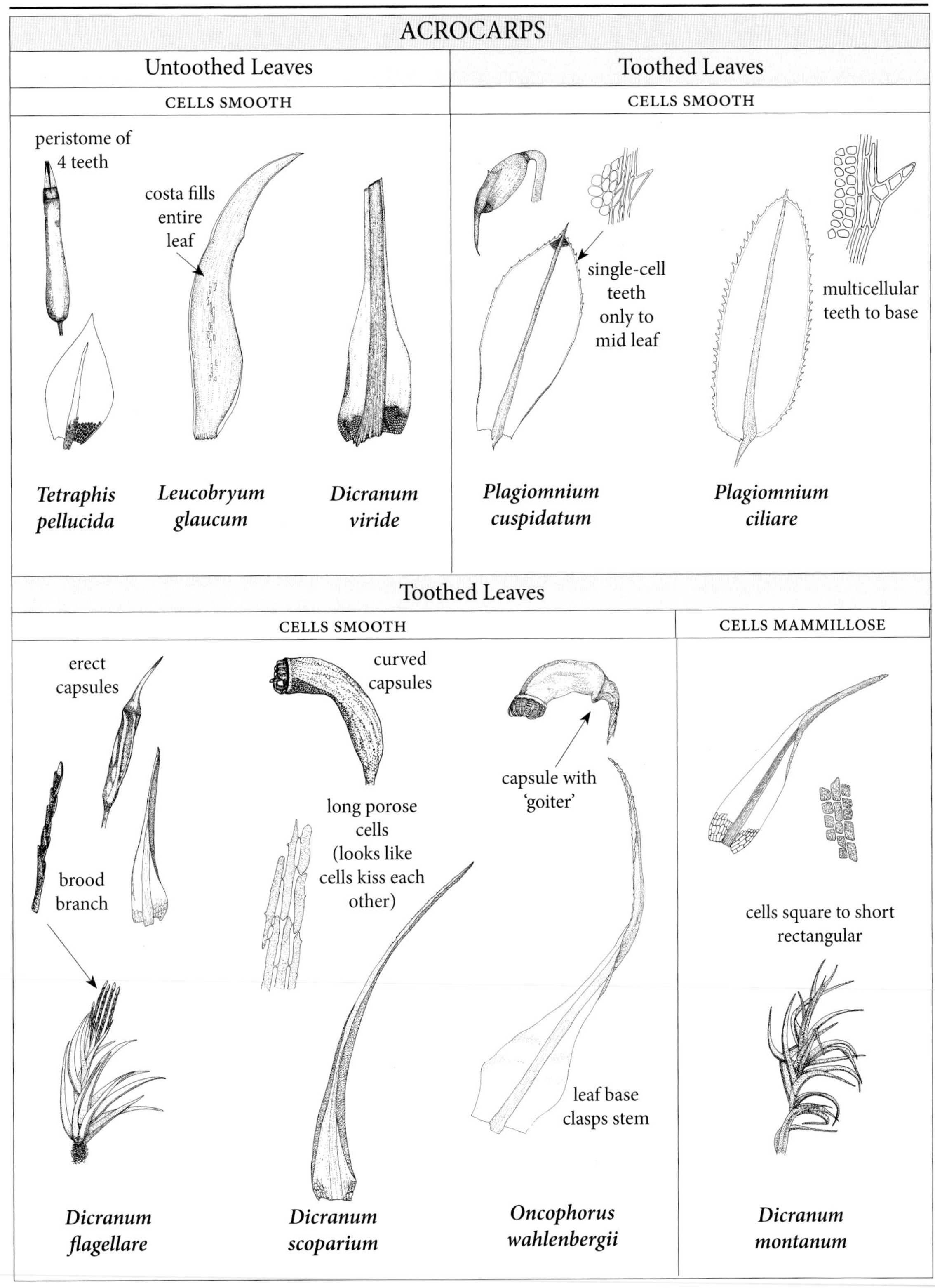
ACROCARPS
Untoothed Leaves
CELLS SMOOTH
peristome of 4 teeth
costa fills entire leaf
Tetraphis pellucida
Leucobryum glaucum
Dicranum viride
Toothed Leaves
CELLS SMOOTH
single-cell teeth only to mid leaf
multicellular teeth to base
Plagiomnium cuspidatum
Plagiomnium ciliare
Toothed Leaves
CELLS SMOOTH
erect capsules
brood branch
curved capsules
long porose cells (looks like cells kiss each other)
capsule with 'goiter'
leaf base clasps stem
Dicranum flagellare
Dicranum scoparium
Oncophorus wahlenbergii
CELLS MAMMILLOSE
cells square to short rectangular
Dicranum montanum

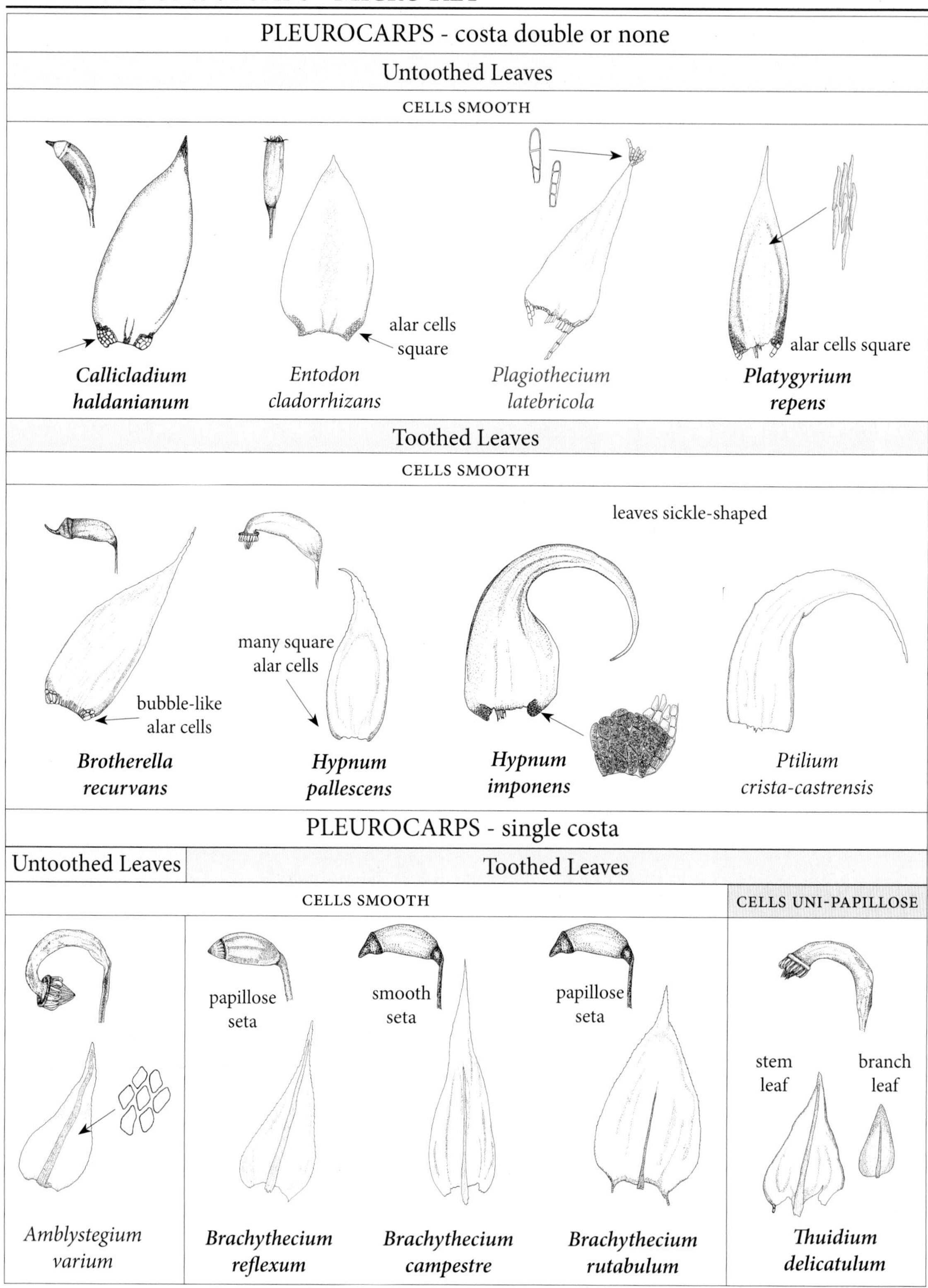
PLEURОCARPS - costa double or none
Untoothed Leaves
CELLS SMOOTH
alar cells
square
alar cells square
Callicladium
haldanianum
Entodon
cladorrhizans
Plagiothecium
latebricola
Platygyrium
repens
Toothed Leaves
CELLS SMOOTH
leaves sickle-shaped
many square
alar cells
bubble-like
alar cells
Brotherella
recurvans
Hypnum
pallescens
Hypnum
imponens
Ptilium
crista-castrensis
PLEUROCARPS - single costa
Untoothed Leaves
Toothed Leaves
CELLS SMOOTH
CELLS UNI-PAPILLOSE
papillose
seta
smooth
seta
papillose
seta
stem
leaf
branch
leaf
Amblystegium
varium
Brachythecium
reflexum
Brachythecium
campestre
Brachythecium
rutabulum
Thuidium
delicatulum

Buxbaumia minakatae

Oncophorus wahlenbergii

Leucobryum glaucum

Tetraphis pellucida

Plagiomnium ciliare

Entodon cladorrhizans

Ptilium crista-castrensis

Brotherella recurvans

Liverworts

Nowellia curvifolia

Lophocolea heterophylla

HUMIC SOIL

(SAW)

If you ever have the chance to visit a boreal forest....you should. Mosses cover virtually 100% of the forest floor. Big mosses too... *Dicranum scoparium, D. polysetum* & *D. ontariense, Hylocomium splendens, Pleurozium schreberi, Hypnum imponens* and *Ptilium crista-castrensis.* They cover not only the ground, but also logs and rocks and tree bases...they are true coveralls.

(SAW)

The hardwood forest floor, however, has hardly anything...oh, mosses are there...but they're up on logs and rocks and tree bases. Why is this? Mosses need two things: light and moisture. The mosses growing in this type of environment are well designed to live in low light levels...but not NO light levels. Falling deciduous leaves are BIG compared with mosses, which quickly get covered in the fall. The only way for them to live is to get up above the falling leaves. Also, growth forms allow the mosses to retain moisture longer once they get wet.

Boreal woods, on the other hand, are dominated by spruce and fir, which have needles that can sift down easily into the large mosses found growing on the ground. Boreal forests also tend to be cooler and moister, which is where mosses (& liverworts) thrive.

Humic soil (which has a lot of organic matter) can be in the open (like in a field that gets mowed every now and then), in the shade under a canopy, along the edge of a stream bank or perhaps where a log has rotted down. Each habitat has a community of bryophytes unique to it given the microclimate of light levels and humidity.

A sampling of mosses on the ground
Boreal Forest Floor
Dicranum scoparium
Dicranum polysetum
Hylocomium splendens
Hypnum imponens
Pleurozium schreberi
Ptilium crista-castrensis
Bazzania trilobata
Ptilidium ciliare
There are so many mosses here!
Same old apple tree in field
Polytrichum commune
Rhytidiadelphus triquetrus
Pleurozium schreberi

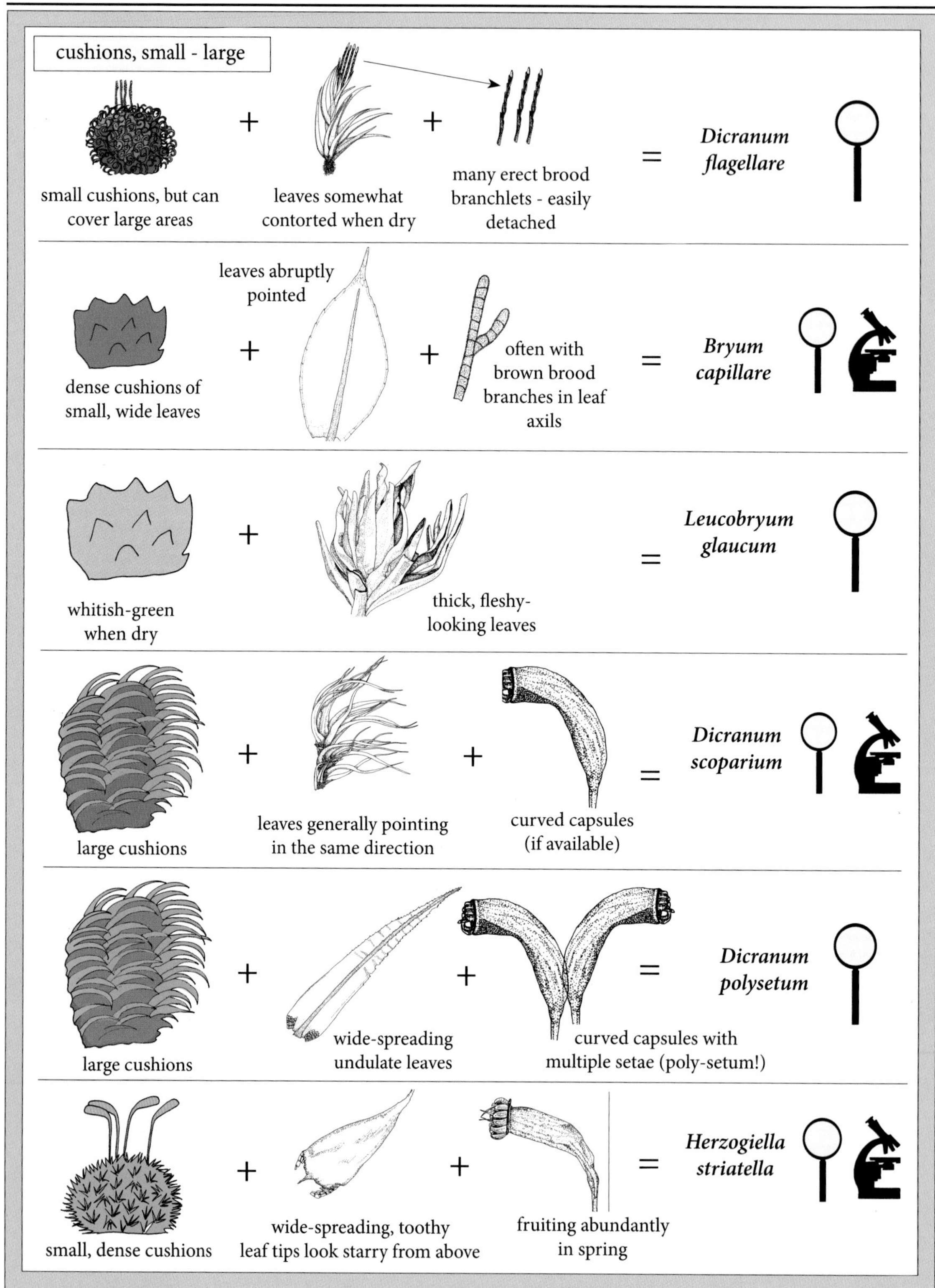
cushions, small - large
small cushions, but can cover large areas
+
leaves somewhat contorted when dry
+
many erect brood branchlets - easily detached
=
Dicranum flagellare
dense cushions of small, wide leaves
+
leaves abruptly pointed
+
often with brown brood branches in leaf axils
=
Bryum capillare
whitish-green when dry
+
thick, fleshy-looking leaves
=
Leucobryum glaucum
large cushions
+
leaves generally pointing in the same direction
+
curved capsules (if available)
=
Dicranum scoparium
large cushions
+
wide-spreading undulate leaves
+
curved capsules with multiple setae (poly-setum!)
=
Dicranum polysetum
small, dense cushions
+
wide-spreading, toothy leaf tips look starry from above
+
fruiting abundantly in spring
=
Herzogiella striatella

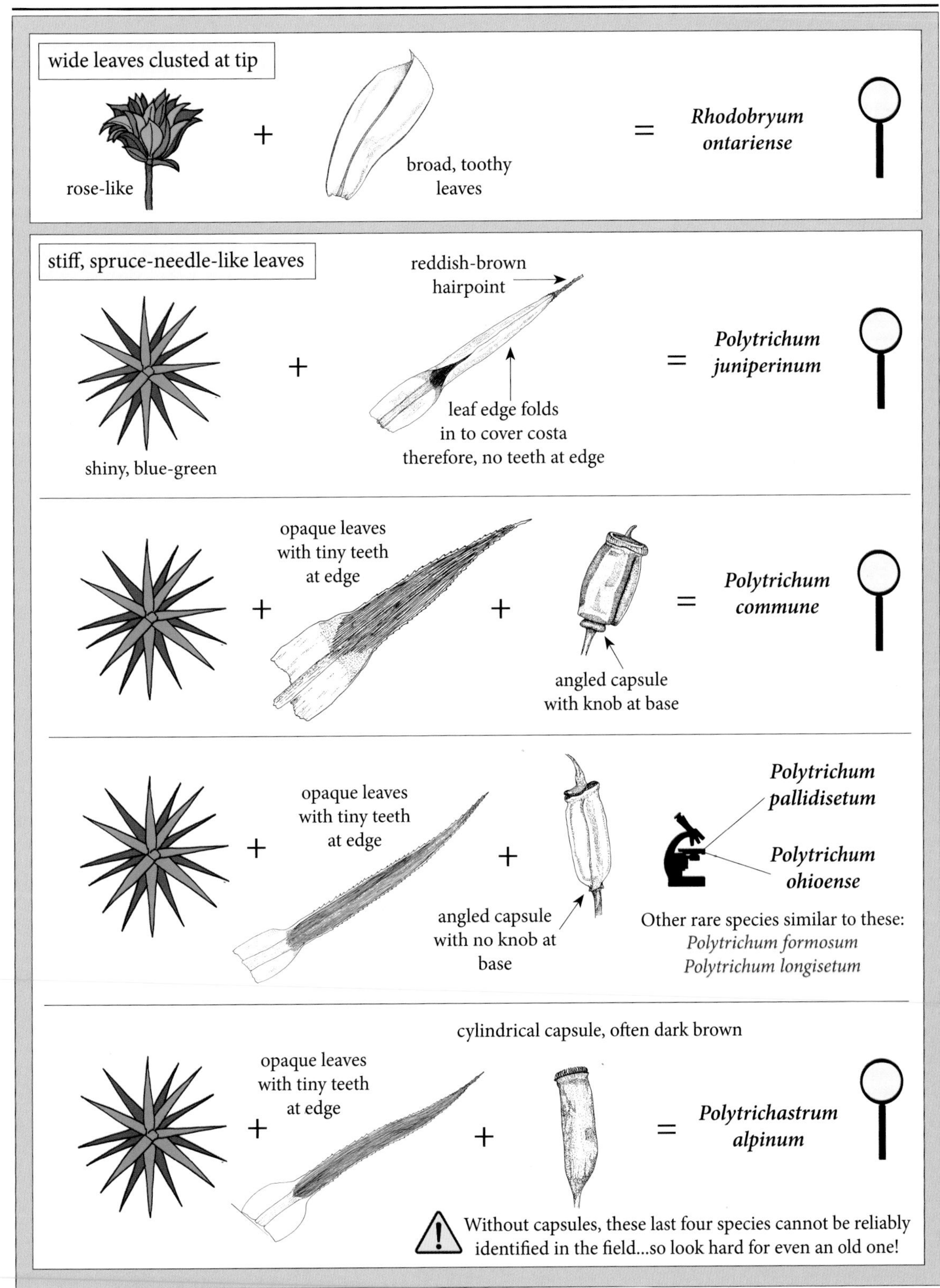
wide leaves clusted at tip
rose-like
broad, toothy leaves
Rhodobryum ontariense
stiff, spruce-needle-like leaves
reddish-brown hairpoint
leaf edge folds in to cover costa therefore, no teeth at edge
shiny, blue-green
Polytrichum juniperinum
opaque leaves with tiny teeth at edge
angled capsule with knob at base
Polytrichum commune
opaque leaves with tiny teeth at edge
angled capsule with no knob at base
Polytrichum pallidisetum
Polytrichum ohioense
Other rare species similar to these:
Polytrichum formosum
Polytrichum longisetum
cylindrical capsule, often dark brown
opaque leaves with tiny teeth at edge
Polytrichastrum alpinum
Without capsules, these last four species cannot be reliably identified in the field...so look hard for even an old one!

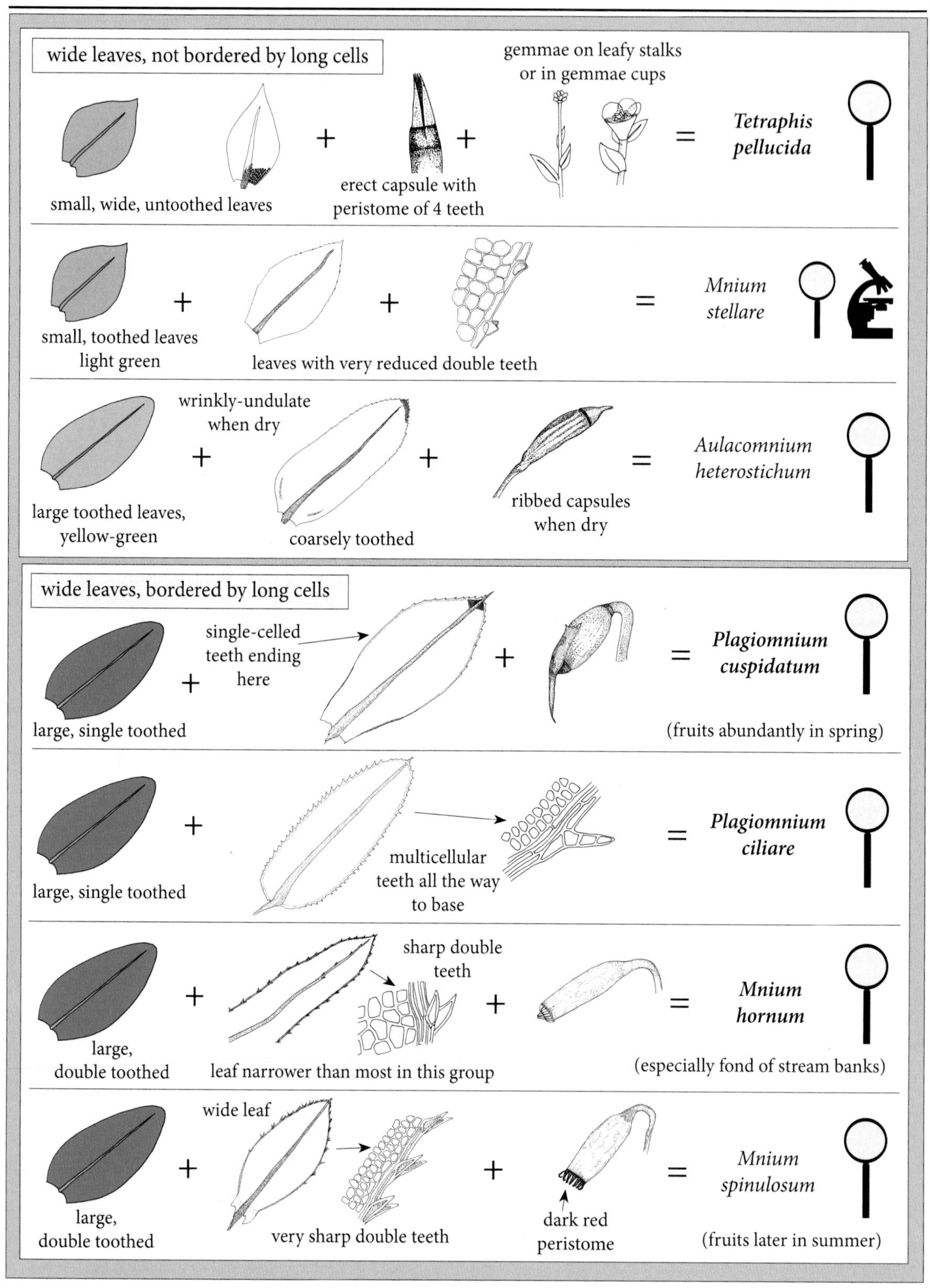
wide leaves, not bordered by long cells
small, wide, untoothed leaves
+
erect capsule with peristome of 4 teeth
+
gemmae on leafy stalks or in gemmae cups
=
Tetraphis pellucida
small, toothed leaves light green
+
leaves with very reduced double teeth
+
=
Mnium stellare
large toothed leaves, yellow-green
+
wrinkly-undulate when dry
coarsely toothed
+
ribbed capsules when dry
=
Aulacomnium heterostichum
wide leaves, bordered by long cells
large, single toothed
+
single-celled teeth ending here
+
=
Plagiomnium cuspidatum
(fruits abundantly in spring)
large, single toothed
+
multicellular teeth all the way to base
=
Plagiomnium ciliare
large, double toothed
+
leaf narrower than most in this group
sharp double teeth
+
=
Mnium hornum
(especially fond of stream banks)
large, double toothed
+
wide leaf
very sharp double teeth
+
dark red peristome
=
Mnium spinulosum
(fruits later in summer)

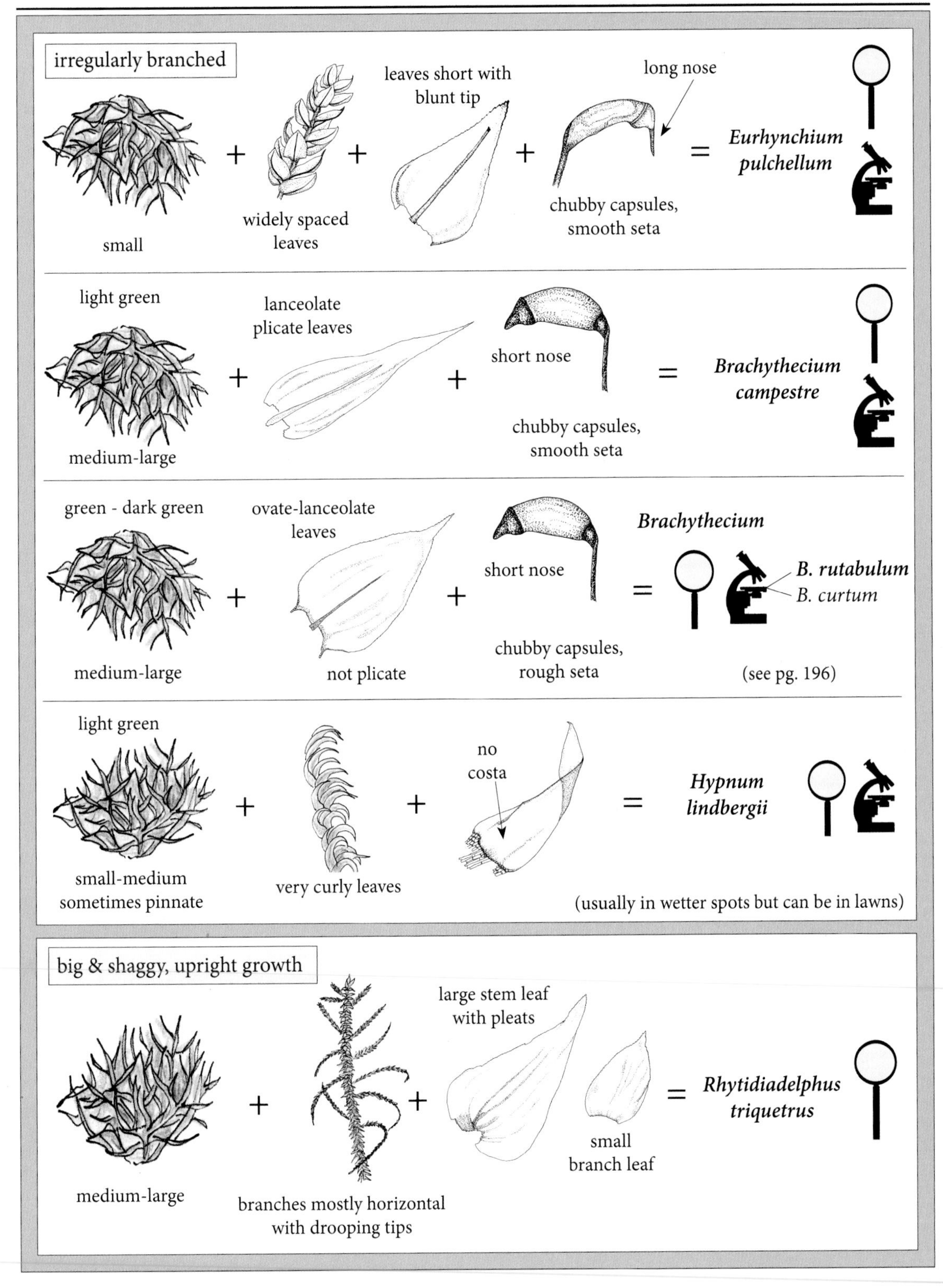
irregularly branched
small
widely spaced leaves
leaves short with blunt tip
long nose
chubby capsules, smooth seta
Eurhynchium pulchellum
light green
medium-large
lanceolate plicate leaves
short nose
chubby capsules, smooth seta
Brachythecium campestre
green - dark green
medium-large
ovate-lanceolate leaves
not plicate
short nose
chubby capsules, rough seta
Brachythecium
B. rutabulum
B. curtum
(see pg. 196)
light green
small-medium sometimes pinnate
very curly leaves
no costa
Hypnum lindbergii
(usually in wetter spots but can be in lawns)
big & shaggy, upright growth
medium-large
branches mostly horizontal with drooping tips
large stem leaf with pleats
small branch leaf
Rhytidiadelphus triquetrus

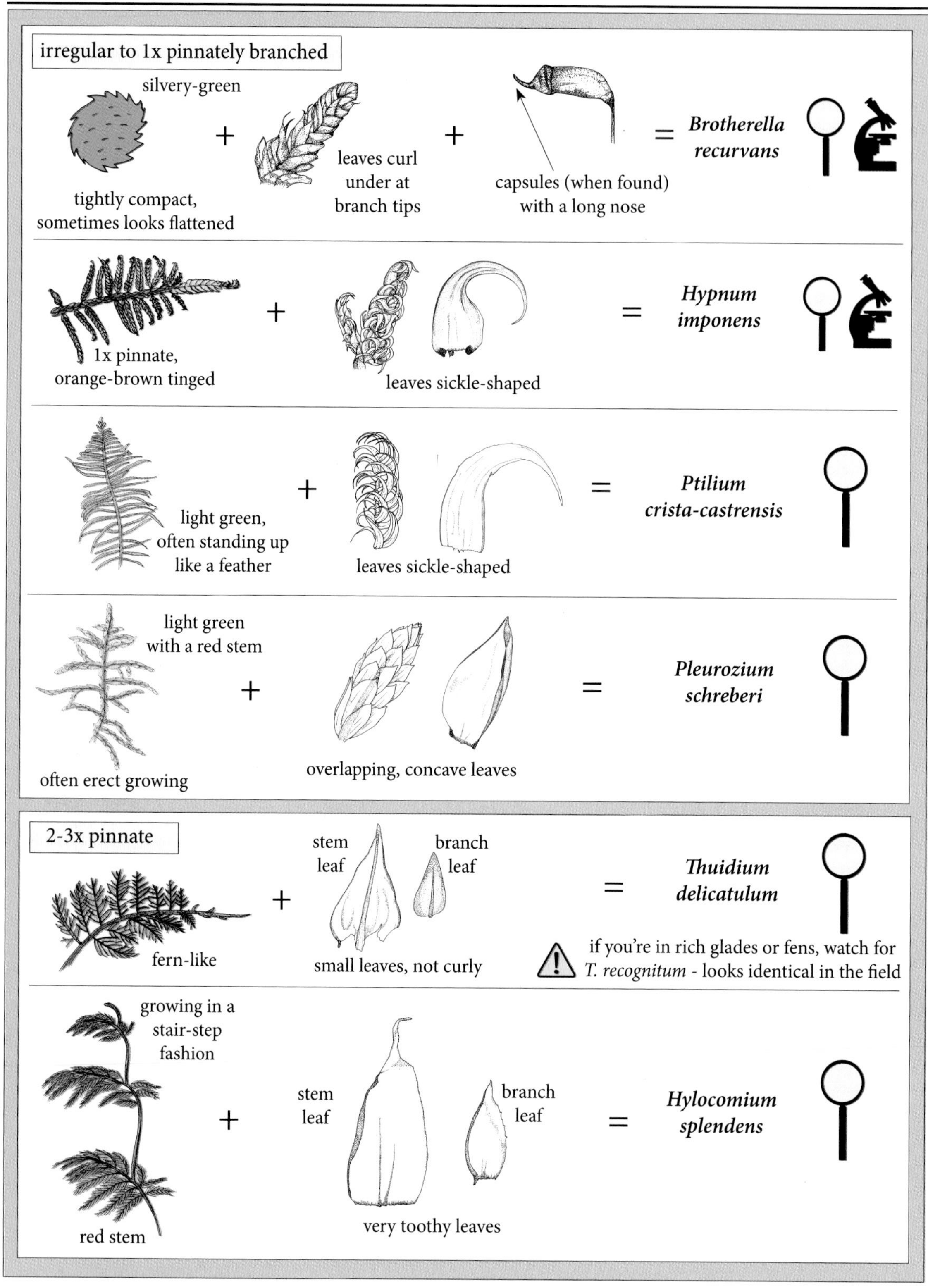
irregular to 1x pinnately branched
silvery-green
tightly compact, sometimes looks flattened
+
leaves curl under at branch tips
+
capsules (when found) with a long nose
=
Brotherella recurvans
1x pinnate, orange-brown tinged
+
leaves sickle-shaped
=
Hypnum imponens
light green, often standing up like a feather
+
leaves sickle-shaped
=
Ptilium crista-castrensis
light green with a red stem
often erect growing
+
overlapping, concave leaves
=
Pleurozium schreberi
2-3x pinnate
fern-like
+
stem leaf
branch leaf
small leaves, not curly
=
Thuidium delicatulum
if you're in rich glades or fens, watch for T. recognitum - looks identical in the field
growing in a stair-step fashion
red stem
+
stem leaf
branch leaf
very toothy leaves
=
Hylocomium splendens

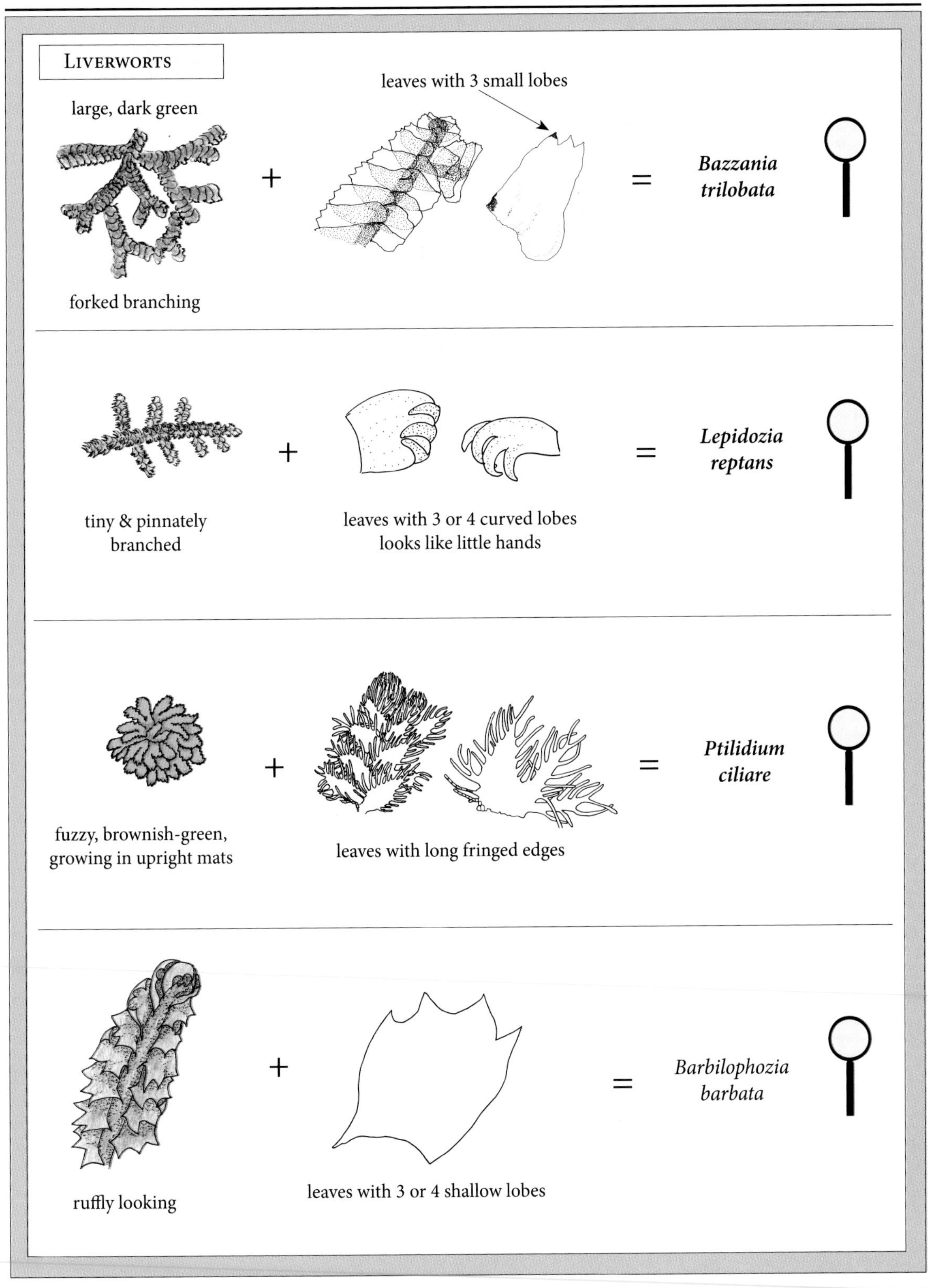
Liverworts
large, dark green
leaves with 3 small lobes
+
=
Bazzania trilobata
forked branching
+
=
Lepidozia reptans
tiny & pinnately branched
leaves with 3 or 4 curved lobes looks like little hands
+
=
Ptilidium ciliare
fuzzy, brownish-green, growing in upright mats
leaves with long fringed edges
+
=
Barbilophozia barbata
ruffly looking
leaves with 3 or 4 shallow lobes

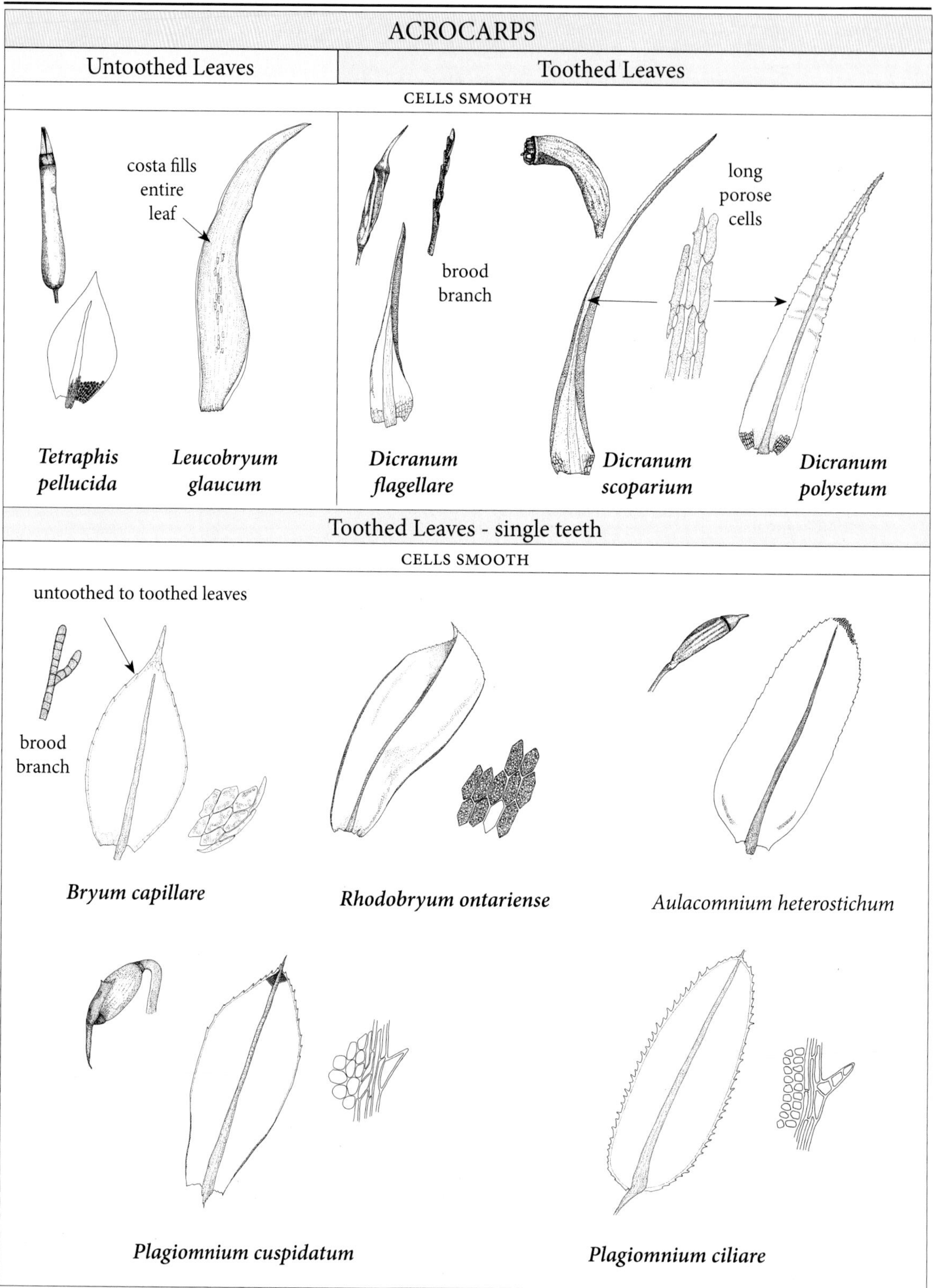
ACROCARPS
Untoothed Leaves
Toothed Leaves
CELLS SMOOTH
costa fills
entire
leaf
brood
branch
long
porose
cells
Tetraphis
pellucida
Leucobryum
glaucum
Dicranum
flagellare
Dicranum
scoparium
Dicranum
polysetum
Toothed Leaves - single teeth
CELLS SMOOTH
untoothed to toothed leaves
brood
branch
Bryum capillare
Rhodobryum ontariense
Aulacomnium heterostichum
Plagiomnium cuspidatum
Plagiomnium ciliare

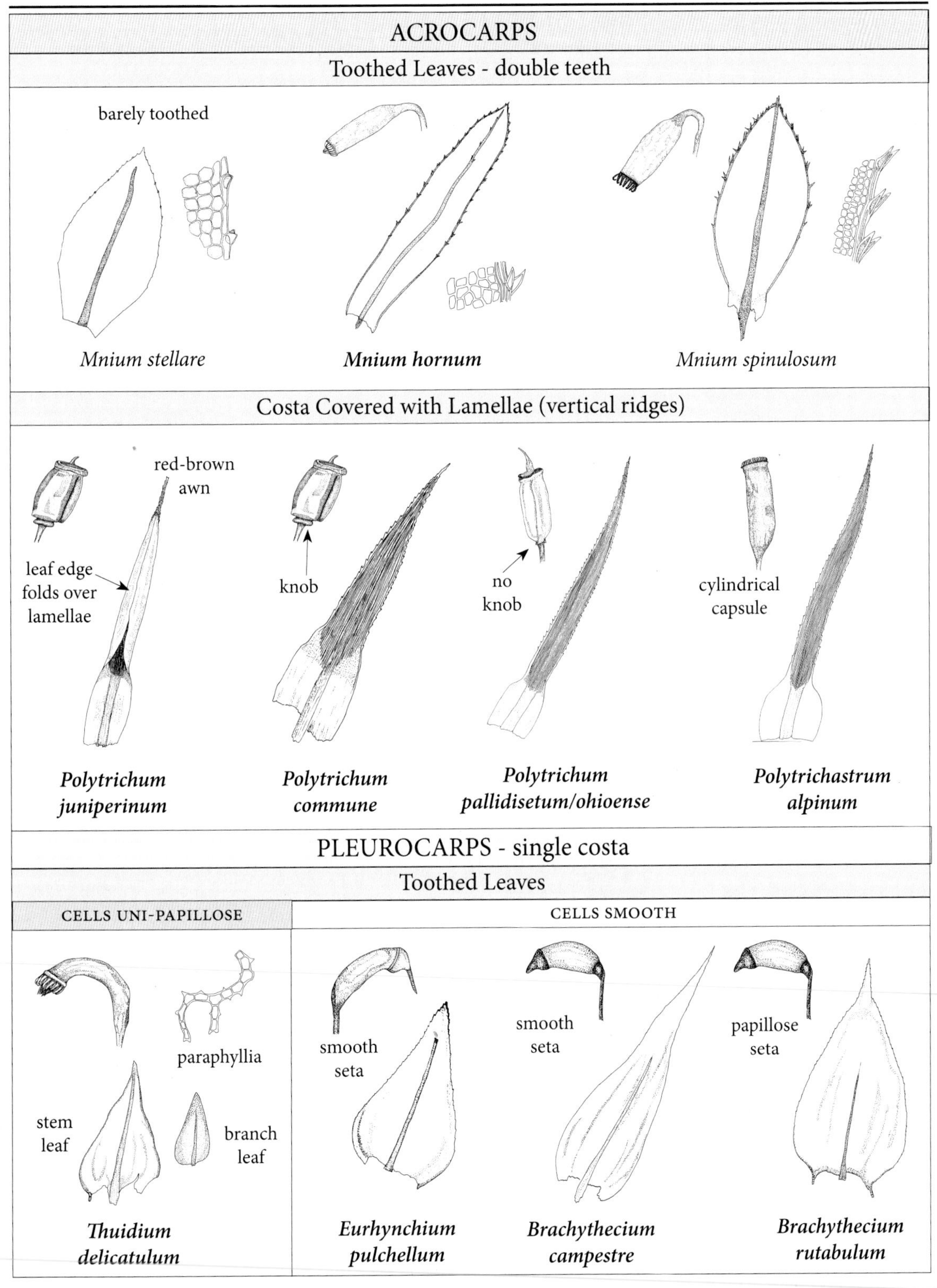
ACROCARPS
Toothed Leaves - double teeth
barely toothed
Mnium stellare
Mnium hornum
Mnium spinulosum
Costa Covered with Lamellae (vertical ridges)
red-brown awn
leaf edge folds over lamellae
knob
no knob
cylindrical capsule
Polytrichum juniperinum
Polytrichum commune
Polytrichum pallidisetum/ohioense
Polytrichastrum alpinum
PLEUROCARPS - single costa
Toothed Leaves
CELLS UNI-PAPILLOSE
CELLS SMOOTH
paraphyllia
stem leaf
branch leaf
smooth seta
smooth seta
papillose seta
Thuidium delicatulum
Eurhynchium pulchellum
Brachythecium campestre
Brachythecium rutabulum

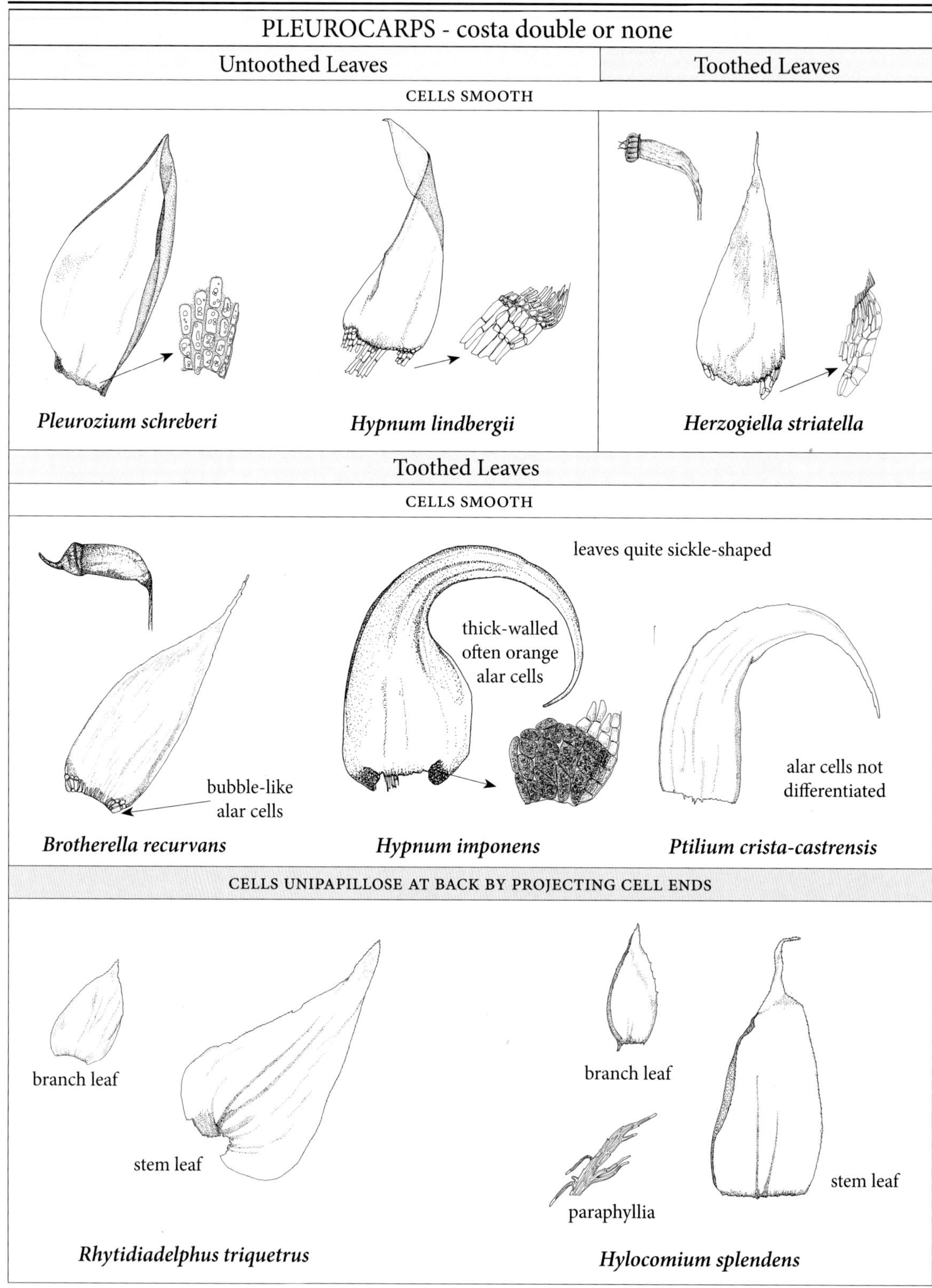
PLEUROCARPS - costa double or none
Untoothed Leaves
Toothed Leaves
CELLS SMOOTH
Pleurozium schreberi
Hypnum lindbergii
Herzogiella striatella
Toothed Leaves
CELLS SMOOTH
leaves quite sickle-shaped
thick-walled
often orange
alar cells
bubble-like
alar cells
alar cells not
differentiated
Brotherella recurvans
Hypnum imponens
Ptilium crista-castrensis
CELLS UNIPAPILLOSE AT BACK BY PROJECTING CELL ENDS
branch leaf
stem leaf
branch leaf
paraphyllia
stem leaf
Rhytidiadelphus triquetrus
Hylocomium splendens

Tetraphis pellucida

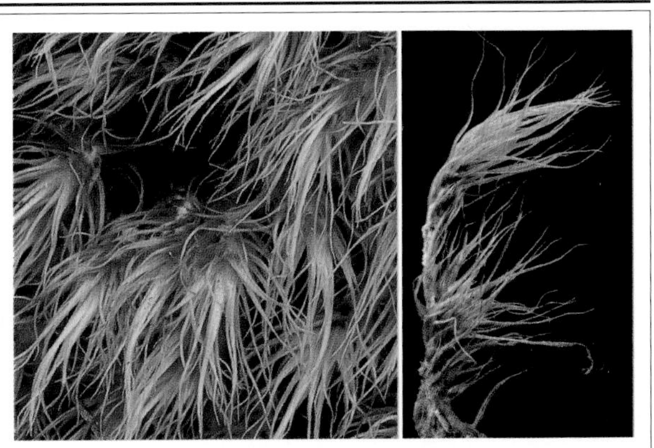

Dicranum scoparium

Aulacomnium heterostichum

Mnium stellare

Dicranum polysetum

Rhytidiadelphus triquetrus

Mnium hornum

Mnium spinulosum

Bryum capillare

Rhodobryum ontariense

Polytrichum commune

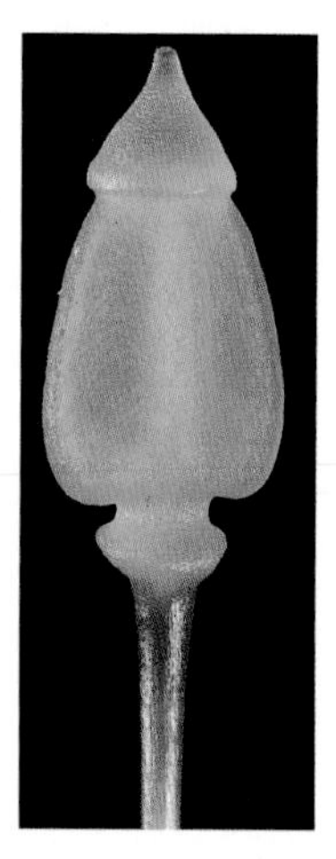

male splash cup
& new capsule

Polytrichum juniperinum & splash cups

Pleurozium schreberi

Hylocomium splendens

LIVERWORTS

Ptilidium ciliare

Lepidozia reptans

ANIMALCULES

Water Bears & Moss Piglets = Tardigrades

One of the most interesting little creatures you might see under the microscope while looking at cells or other features is the tardigrade. They have plump little bodies with eight legs with claws (the hind ones face backwards and are used mostly for attaching to things). Tardigrades are really hardy and can withstand environments as cold as -328° F (-200° C) or highs of more than 300° F (148.9 C). They can also survive radiation, boiling liquids, as well as massive amounts of pressure! In many of these extreme conditions, they survive by going into a dehydrated ball, called a tun, by retracting their head and legs. But just add water and the tardigrade can come back to life in just a few hours! Tardigrades were discovered by a German pastor, Johann August Ephraim Goeze, in 1773. He named them Tardigrada, which means "slow stepper."

Wheeled Animalcules = Rotifers

These are seen more often than water bears. They especially love to hang out in the little water sacs on the back of the liverwort *Frullania*. If you put a branch of *Frullania* under the microscope with the back side up (so you are able to see the sacs easily) and scan around, you might see little creatures poking their heads in and out of the sacs. They have cilia around the head part (called a corona), and the movement of the cilia makes it look like little wheels are spinning - hence the old name of wheeled animalcules. The purpose of the rotary motion of the cilia is to create currents in the surrounding water to bring minute organisms to their mouth, as well as for propulsion. There are many types of rotifers, but the ones commonly seen in *Frullania* are cylindrical. Besides the head they have a foot that anchors the rotifer to its substrate.

Like tardigrades, some rotifers also can resist drying out by secreting gelatinous envelopes which prevent further drying.

Both of these microscopic creatures are fun to find and watch and make a fun distraction from identifying moss!

MINERAL SOIL

Mineral soil is soil that contains little or no organic material. It can be gravelly and open, like this bank near a reservoir, or can be clay-like with open patches, as in the photo below in a garden. The mosses that grow in these types of places are often colonizers, or fast growing annuals that come in, grow quickly and produce many spores. The substrates are often unstable, and the more perennial mosses or plants have a hard time getting established. It is also a very extreme place to live, with wide swings in temperatures and moisture levels, and the mosses that live here need to be able to cope with these conditions.

Physcomitrium pyriforme in an open area

Other mosses and liverworts grow on mineral soil in the shade. These would be places where erosion has taken place, such as steep banks, trail edges or tip-up mounds in the woods...also unstable habitats. The mosses that inhabit both these situations, both open and shaded, avoid competition by means of colonizing and reproducing quickly. Sometimes in shadier places mosses can prevent erosion and start the process of building up the soil for other more perennial mosses and plants.

Open areas where you can find these bryophytes include gravel pits, roadsides, spaces next to buildings, railroad beds and access areas to powerline cuts. Where the ground remains wetter for much of the time, look at the section on wetland specialists.

New sporophytes of *Pogonatum pensilvanicum & Atrichum angustatum* growing on open soil on bank in woods

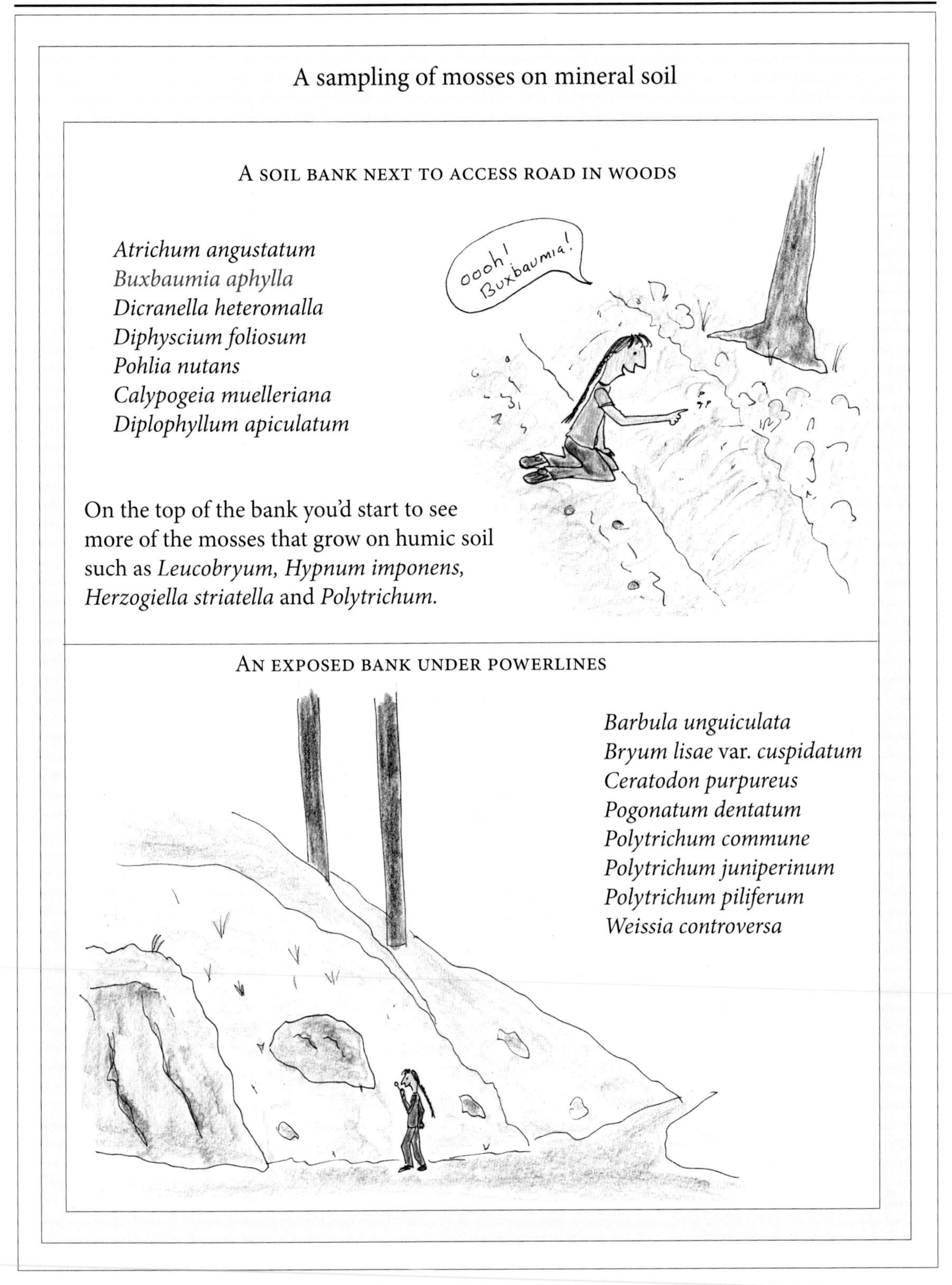
A sampling of mosses on mineral soil
A soil bank next to access road in woods
Atrichum angustatum
Buxbaumia aphylla
Dicranella heteromalla
Diphyscium foliosum
Pohlia nutans
Calypogeia muelleriana
Diplophyllum apiculatum
oooh! Buxbaumia!
On the top of the bank you'd start to see more of the mosses that grow on humic soil such as *Leucobryum, Hypnum imponens, Herzogiella striatella* and *Polytrichum.*
An exposed bank under powerlines
Barbula unguiculata
Bryum lisae var. *cuspidatum*
Ceratodon purpureus
Pogonatum dentatum
Polytrichum commune
Polytrichum juniperinum
Polytrichum piliferum
Weissia controversa

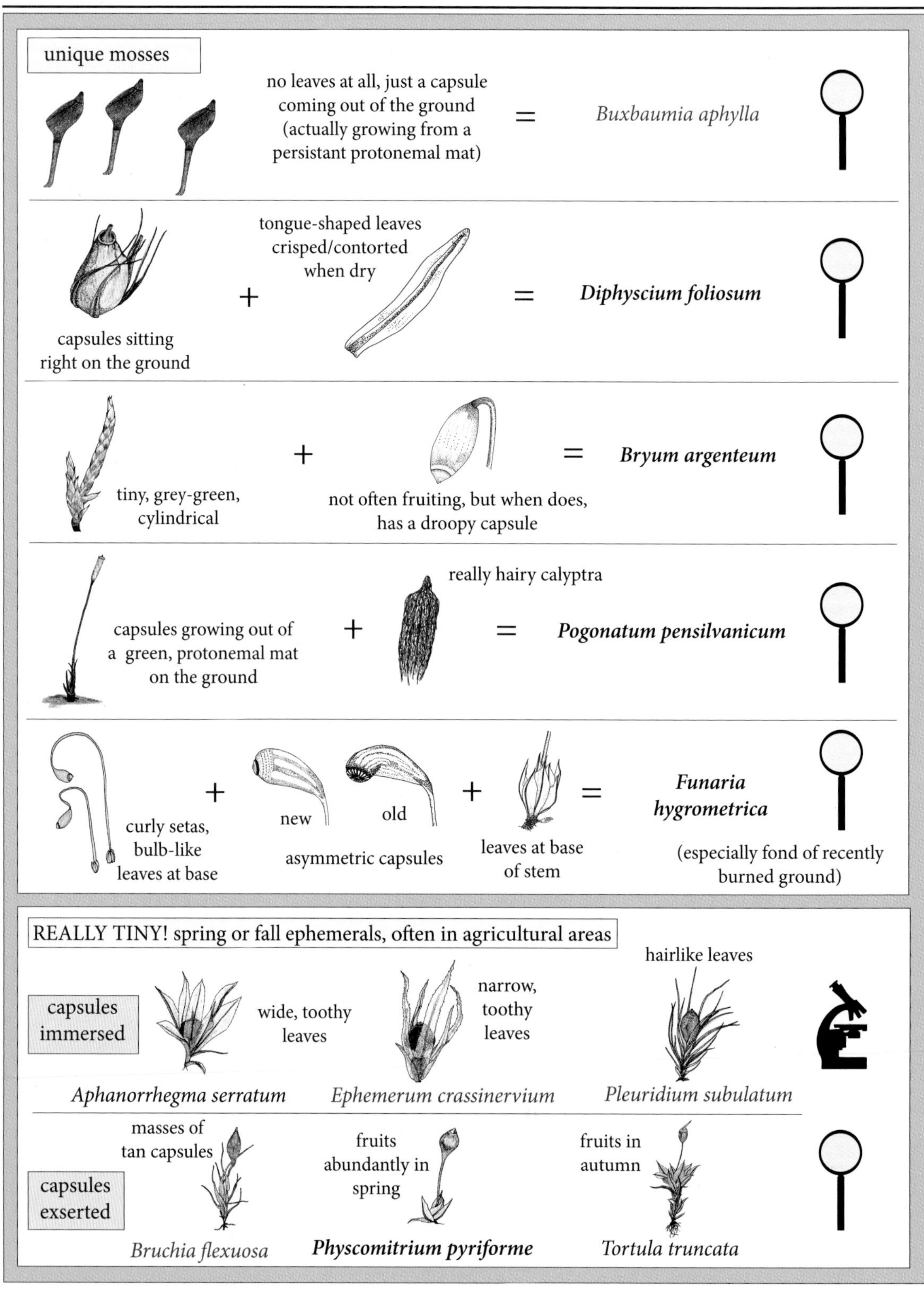

unique mosses
no leaves at all, just a capsule coming out of the ground (actually growing from a persistant protonemal mat)
=
Buxbaumia aphylla
capsules sitting right on the ground
+
tongue-shaped leaves crisped/contorted when dry
=
Diphyscium foliosum
tiny, grey-green, cylindrical
+
not often fruiting, but when does, has a droopy capsule
=
Bryum argenteum
capsules growing out of a green, protonemal mat on the ground
+
really hairy calyptra
=
Pogonatum pensilvanicum
curly setas, bulb-like leaves at base
+
new
old
asymmetric capsules
+
leaves at base of stem
=
Funaria hygrometrica
(especially fond of recently burned ground)
REALLY TINY! spring or fall ephemerals, often in agricultural areas
capsules immersed
wide, toothy leaves
Aphanorrhegma serratum
narrow, toothy leaves
Ephemerum crassinervium
hairlike leaves
Pleuridium subulatum
capsules exserted
masses of tan capsules
Bruchia flexuosa
fruits abundantly in spring
Physcomitrium pyriforme
fruits in autumn
Tortula truncata

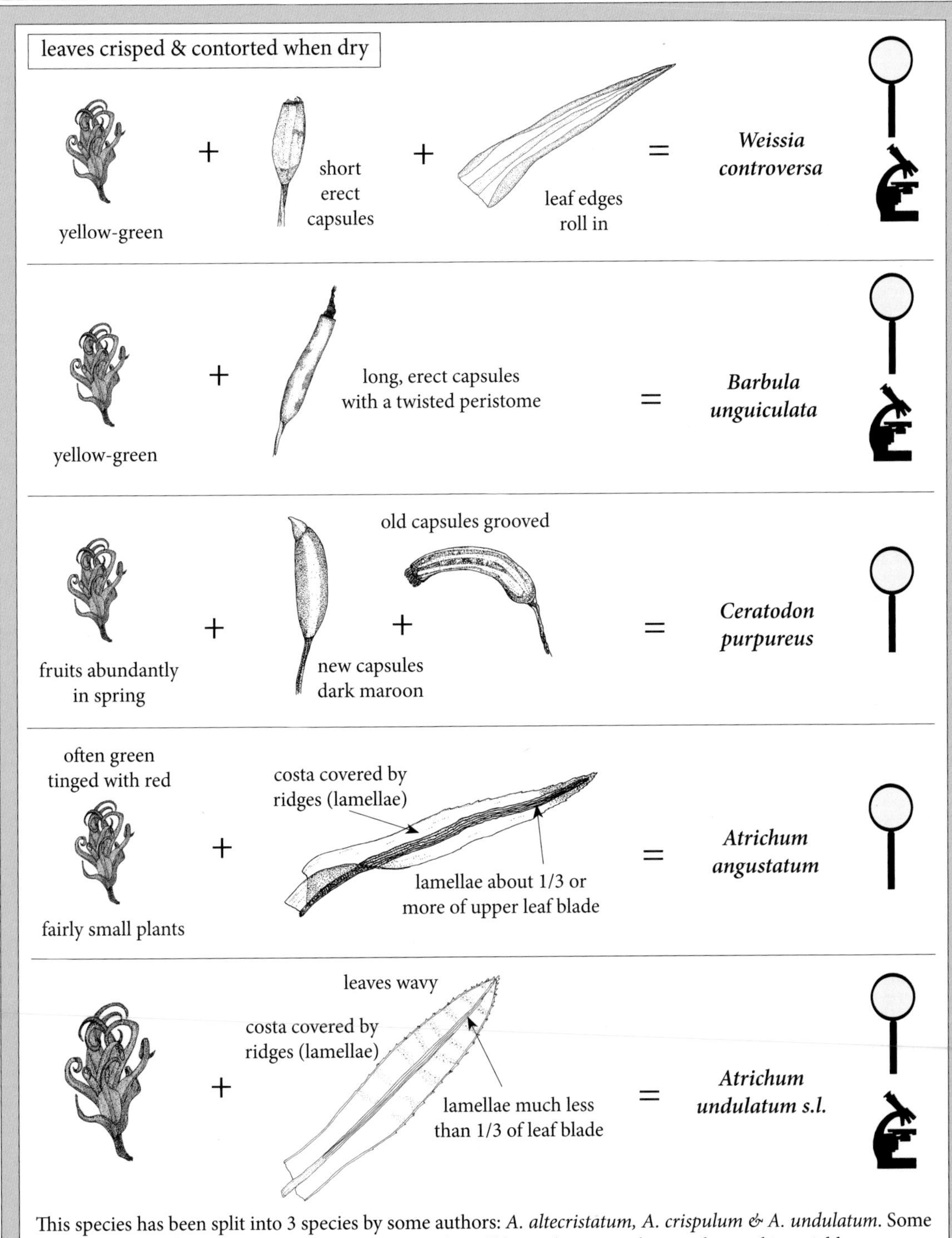

This species has been split into 3 species by some authors: *A. altecristatum, A. crispulum & A. undulatum*. Some say *A. undulatum* is not found on the East Coast. They all have characters that overlap, and invariably you come up within the overlap zone or they have characters that fit one species & other characters that fit another. It's been left here in the broadest sense (*s.l. = sensu lato*). See page 184 for more information.

Mineral Soil - Quick & Dirty Field Key

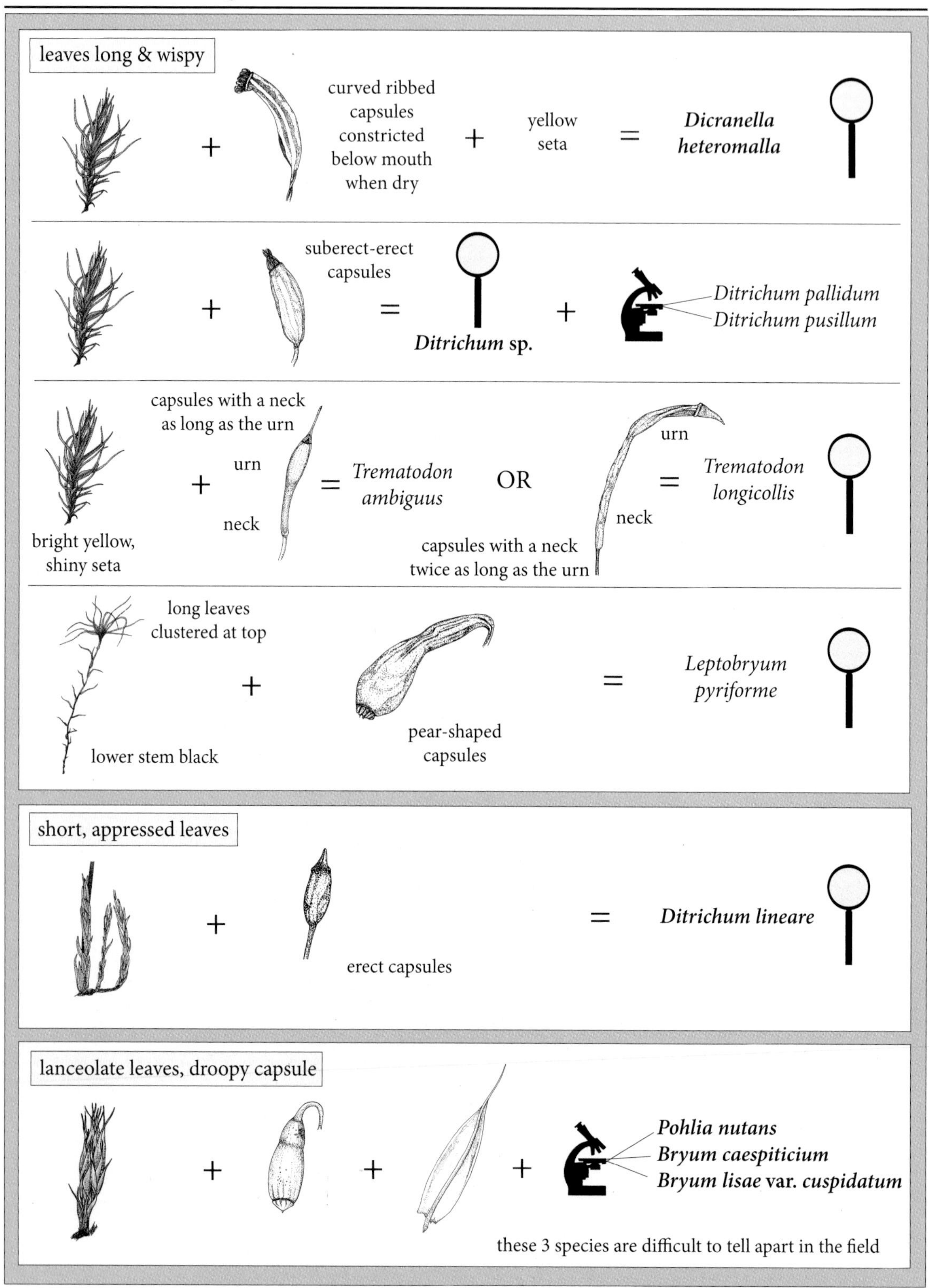

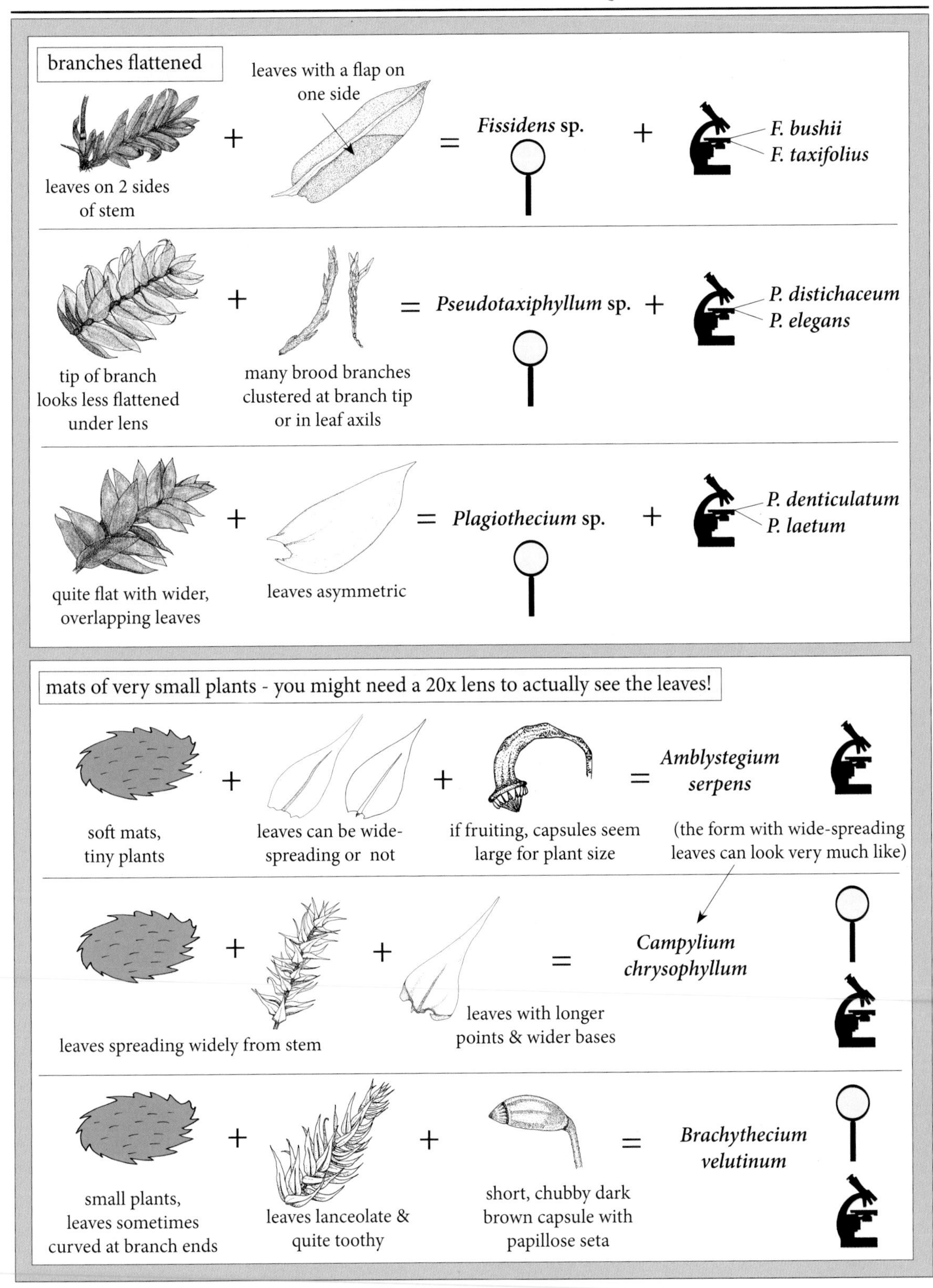

branches flattened
leaves on 2 sides of stem
leaves with a flap on one side
Fissidens sp.
F. bushii
F. taxifolius
tip of branch looks less flattened under lens
many brood branches clustered at branch tip or in leaf axils
Pseudotaxiphyllum sp.
P. distichaceum
P. elegans
quite flat with wider, overlapping leaves
leaves asymmetric
Plagiothecium sp.
P. denticulatum
P. laetum
mats of very small plants - you might need a 20x lens to actually see the leaves!
soft mats, tiny plants
leaves can be wide-spreading or not
if fruiting, capsules seem large for plant size
Amblystegium serpens
(the form with wide-spreading leaves can look very much like)
leaves spreading widely from stem
leaves with longer points & wider bases
Campylium chrysophyllum
small plants, leaves sometimes curved at branch ends
leaves lanceolate & quite toothy
short, chubby dark brown capsule with papillose seta
Brachythecium velutinum

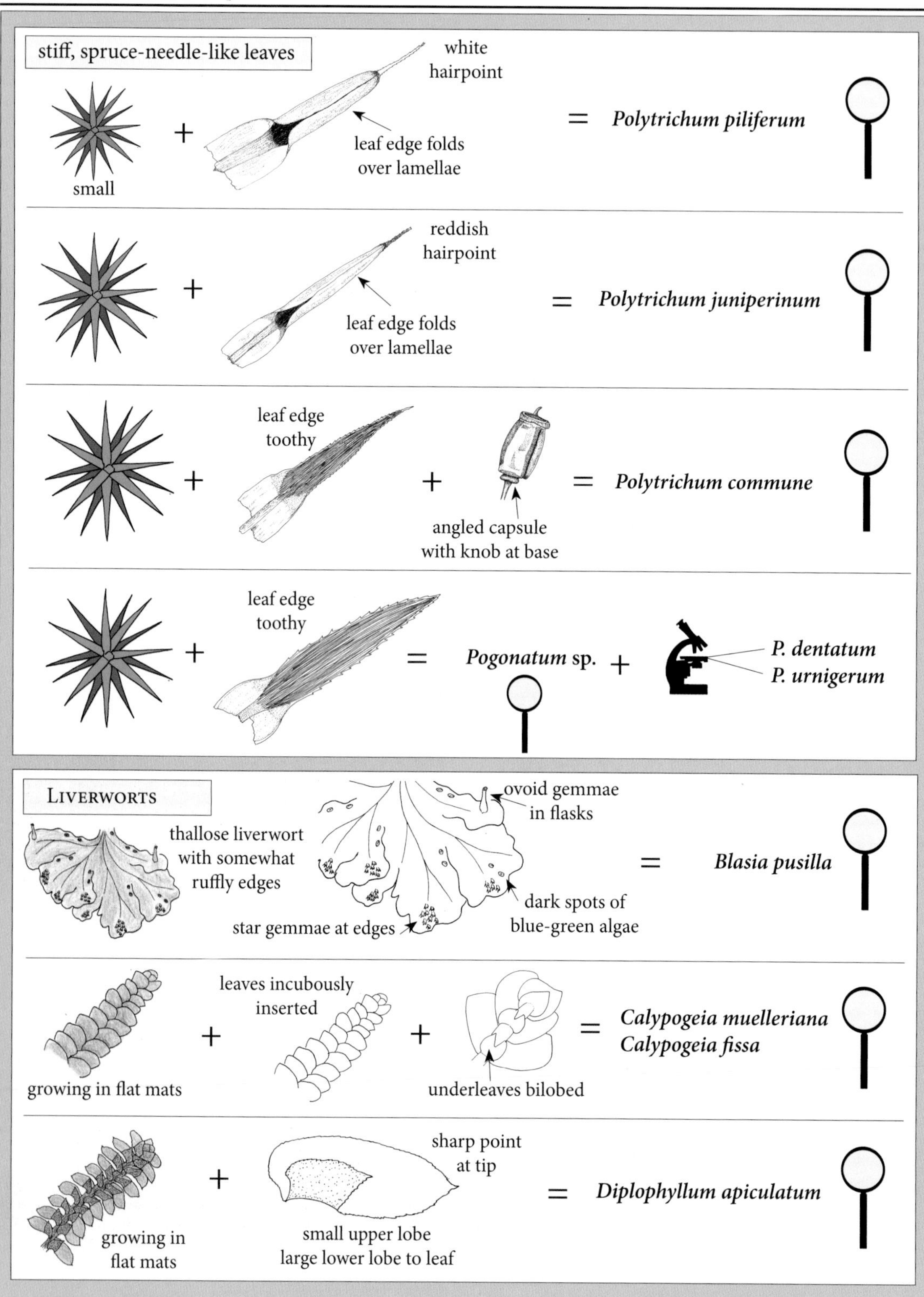
stiff, spruce-needle-like leaves
white hairpoint
leaf edge folds over lamellae
small
= Polytrichum piliferum
reddish hairpoint
leaf edge folds over lamellae
= Polytrichum juniperinum
leaf edge toothy
angled capsule with knob at base
= Polytrichum commune
leaf edge toothy
= Pogonatum sp. +
P. dentatum
P. urnigerum
Liverworts
thallose liverwort with somewhat ruffly edges
ovoid gemmae in flasks
star gemmae at edges
dark spots of blue-green algae
= Blasia pusilla
growing in flat mats
leaves incubously inserted
underleaves bilobed
= Calypogeia muelleriana
Calypogeia fissa
growing in flat mats
sharp point at tip
small upper lobe large lower lobe to leaf
= Diplophyllum apiculatum

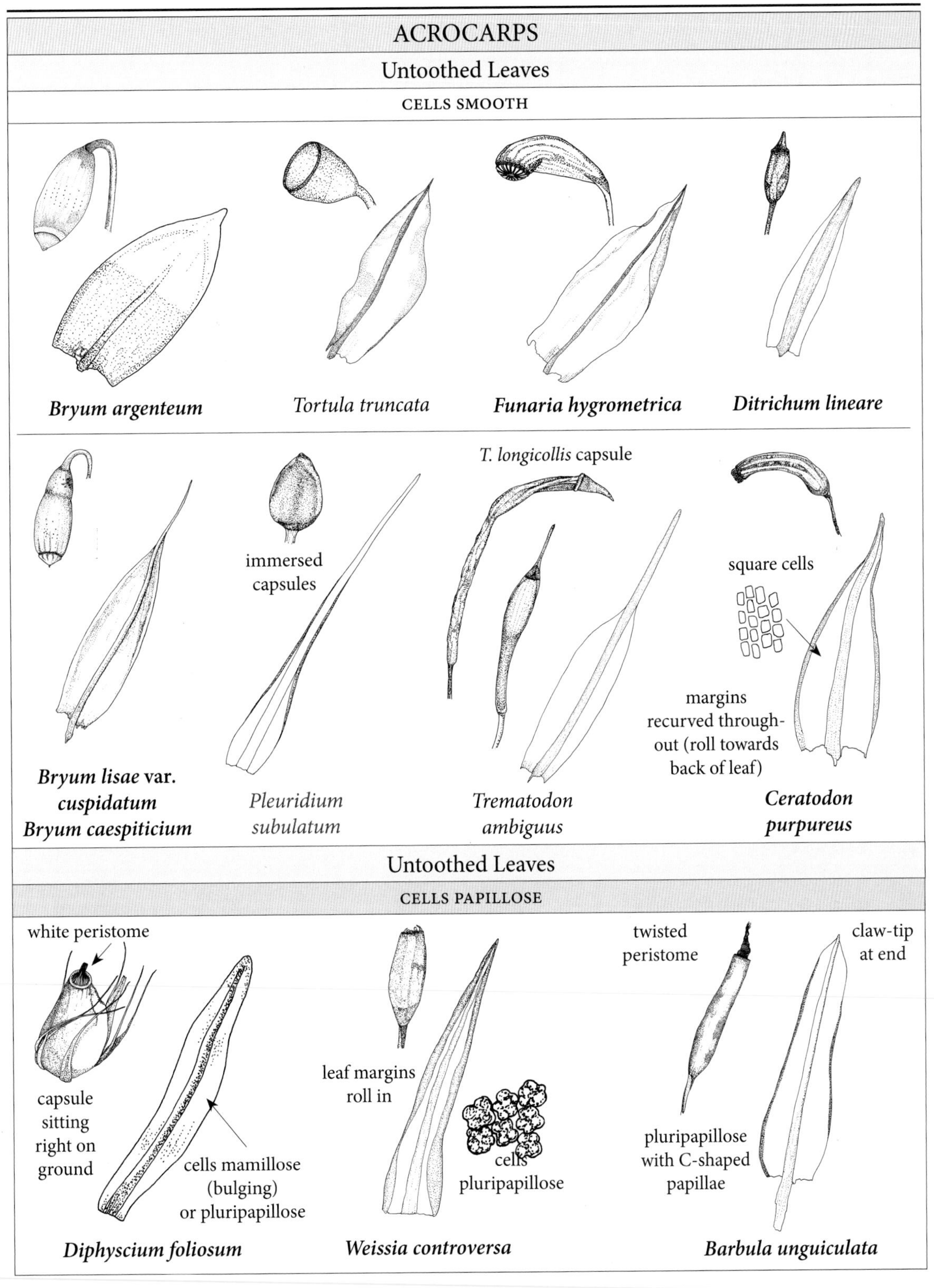
ACROCARPS
Untoothed Leaves
CELLS SMOOTH
Bryum argenteum
Tortula truncata
Funaria hygrometrica
Ditrichum lineare
immersed capsules
T. longicollis capsule
square cells
margins recurved throughout (roll towards back of leaf)
Bryum lisae var. cuspidatum
Bryum caespiticium
Pleuridium subulatum
Trematodon ambiguus
Ceratodon purpureus
Untoothed Leaves
CELLS PAPILLOSE
white peristome
capsule sitting right on ground
cells mamillose (bulging) or pluripapillose
leaf margins roll in
cells pluripapillose
twisted peristome
claw-tip at end
pluripapillose with C-shaped papillae
Diphyscium foliosum
Weissia controversa
Barbula unguiculata

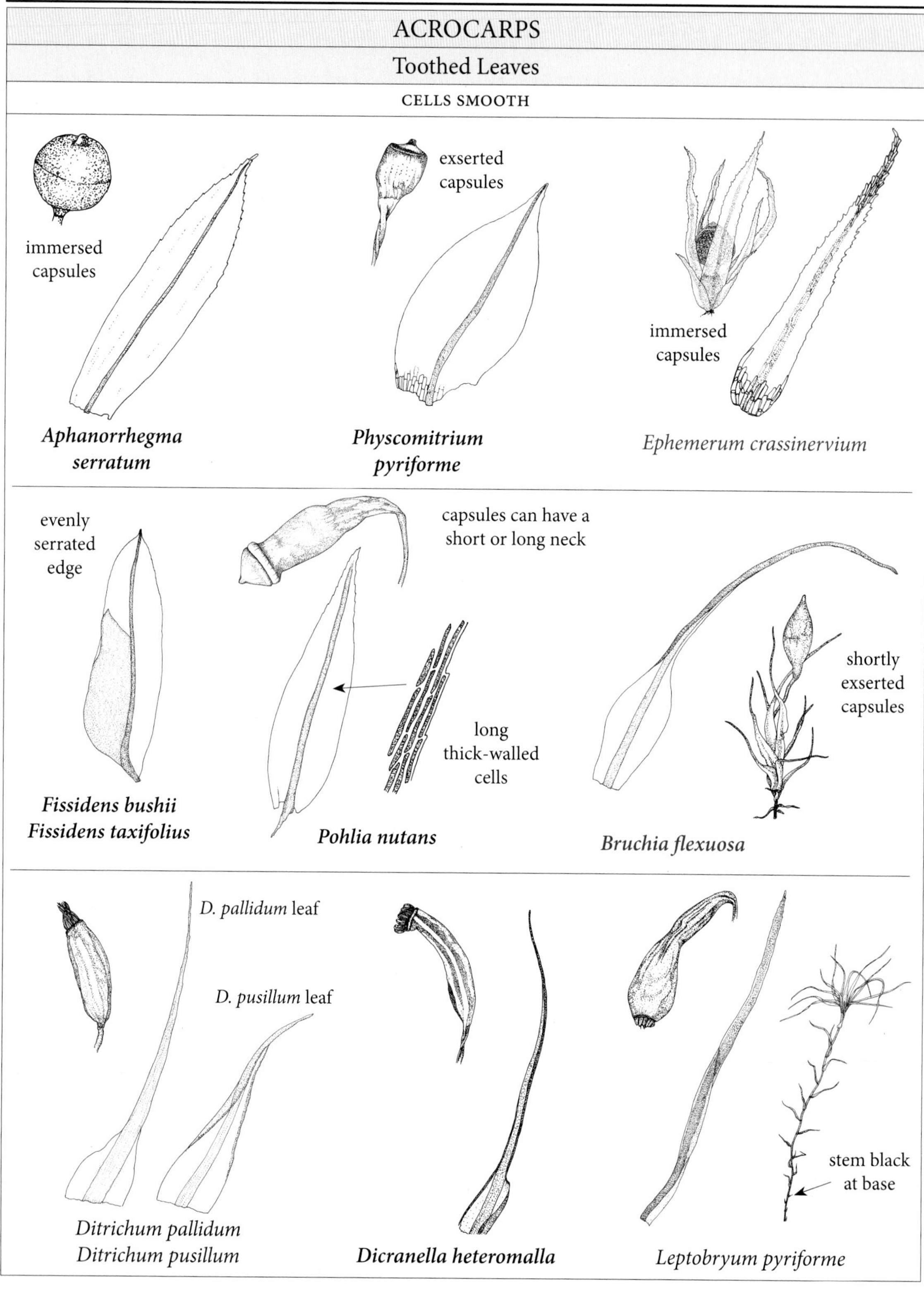
ACROCARPS
Toothed Leaves
CELLS SMOOTH
immersed capsules
Aphanorrhegma serratum
exserted capsules
Physcomitrium pyriforme
immersed capsules
Ephemerum crassinervium
evenly serrated edge
Fissidens bushii
Fissidens taxifolius
capsules can have a short or long neck
long thick-walled cells
Pohlia nutans
shortly exserted capsules
Bruchia flexuosa
D. pallidum leaf
D. pusillum leaf
Ditrichum pallidum
Ditrichum pusillum
Dicranella heteromalla
stem black at base
Leptobryum pyriforme

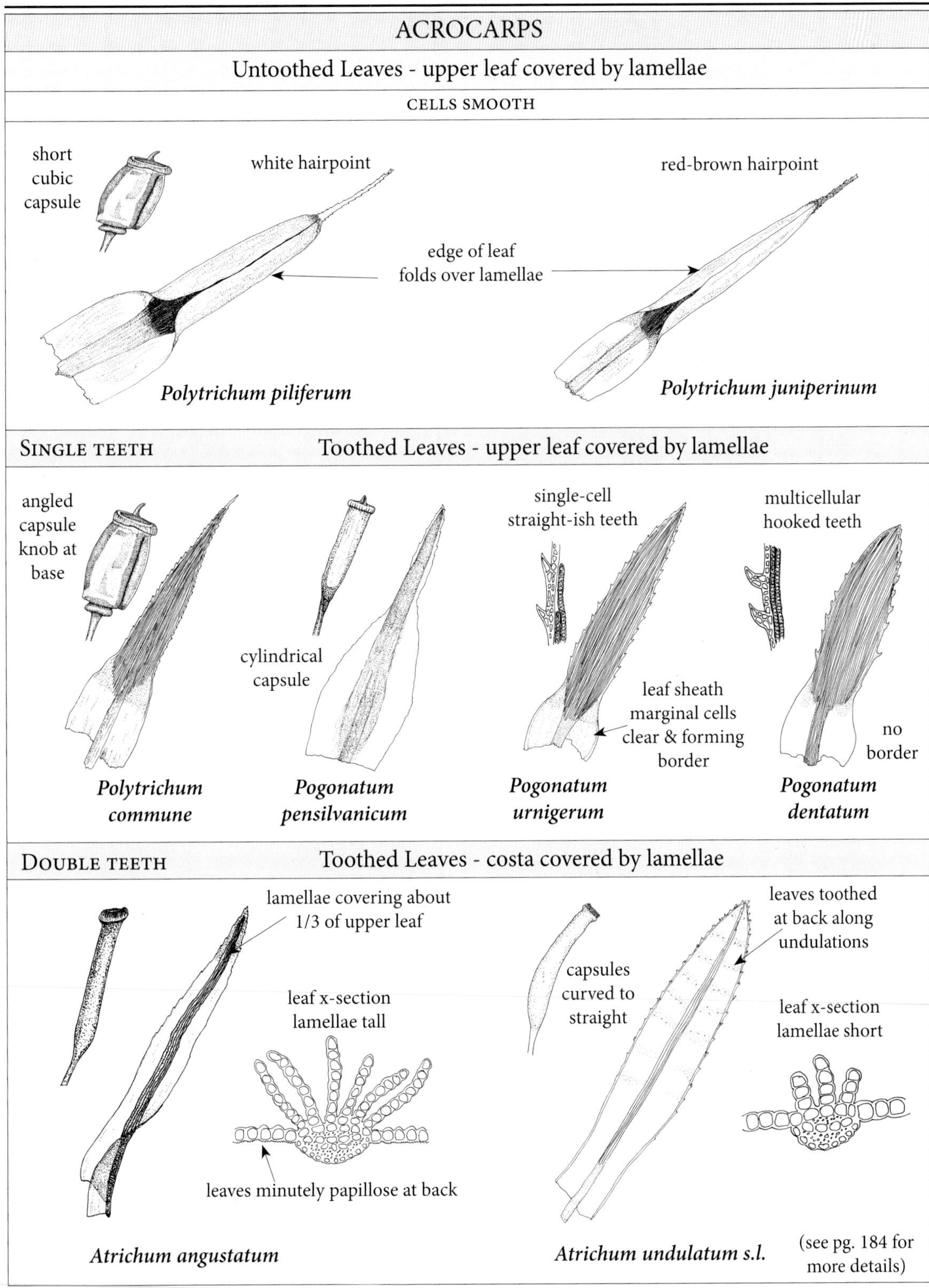
ACROCARPS
Untoothed Leaves - upper leaf covered by lamellae
CELLS SMOOTH
short cubic capsule
white hairpoint
red-brown hairpoint
edge of leaf folds over lamellae
Polytrichum piliferum
Polytrichum juniperinum
SINGLE TEETH
Toothed Leaves - upper leaf covered by lamellae
angled capsule knob at base
cylindrical capsule
single-cell straight-ish teeth
multicellular hooked teeth
leaf sheath marginal cells clear & forming border
no border
Polytrichum commune
Pogonatum pensilvanicum
Pogonatum urnigerum
Pogonatum dentatum
DOUBLE TEETH
Toothed Leaves - costa covered by lamellae
lamellae covering about 1/3 of upper leaf
leaf x-section lamellae tall
leaves minutely papillose at back
capsules curved to straight
leaves toothed at back along undulations
leaf x-section lamellae short
Atrichum angustatum
Atrichum undulatum s.l.
(see pg. 184 for more details)

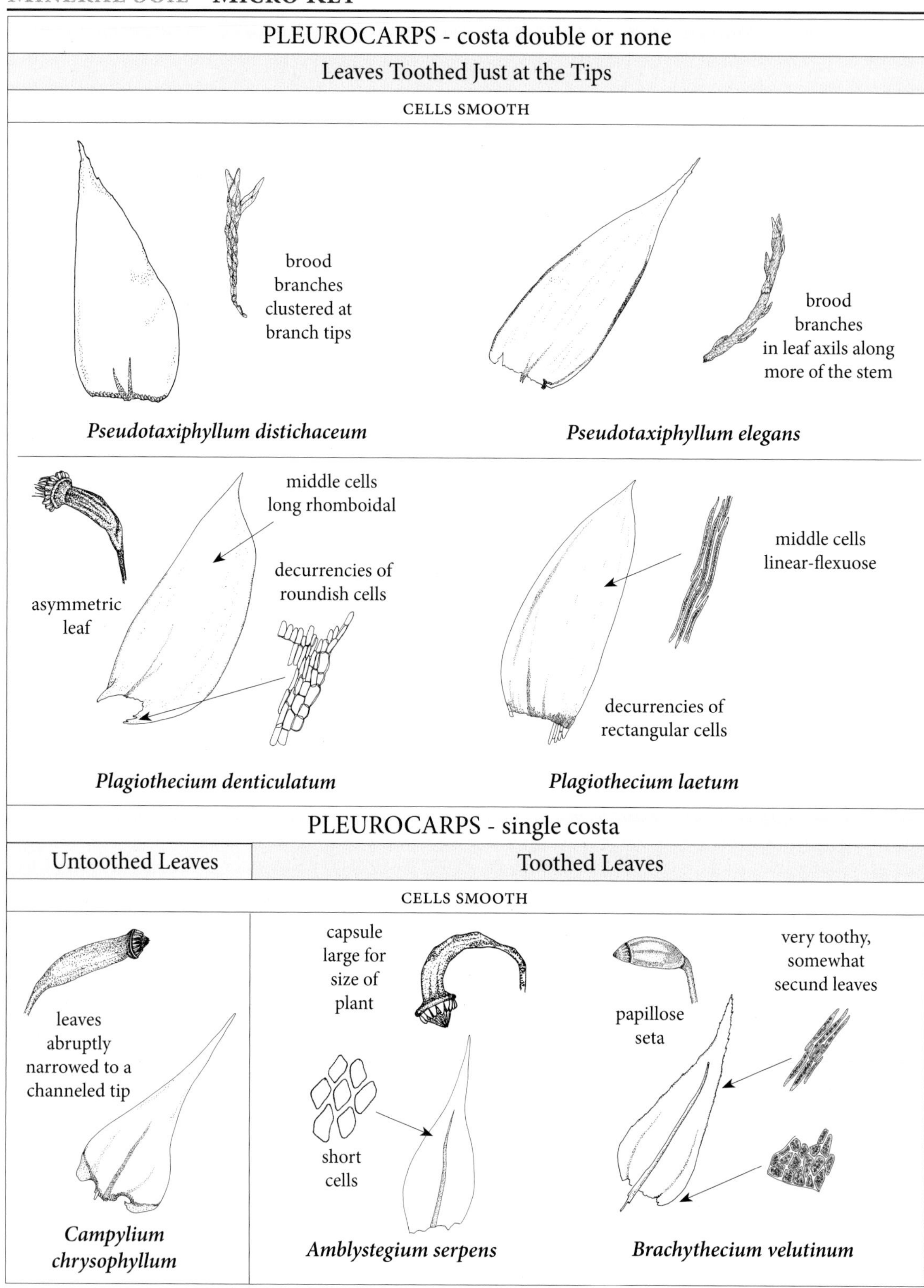
PLEUROCARPS - costa double or none
Leaves Toothed Just at the Tips
CELLS SMOOTH
brood branches clustered at branch tips
Pseudotaxiphyllum distichaceum
brood branches in leaf axils along more of the stem
Pseudotaxiphyllum elegans
middle cells long rhomboidal
decurrencies of roundish cells
asymmetric leaf
Plagiothecium denticulatum
middle cells linear-flexuose
decurrencies of rectangular cells
Plagiothecium laetum
PLEUROCARPS - single costa
Untoothed Leaves
Toothed Leaves
CELLS SMOOTH
leaves abruptly narrowed to a channeled tip
Campylium chrysophyllum
capsule large for size of plant
short cells
Amblystegium serpens
papillose seta
very toothy, somewhat secund leaves
Brachythecium velutinum

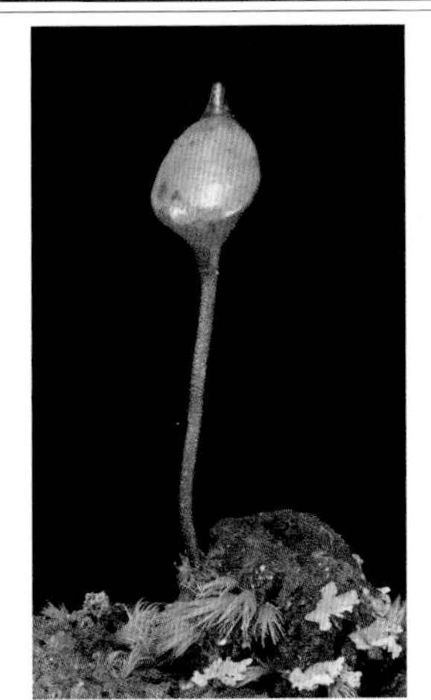

Buxbaumia aphylla

Diphyscium foliosum

Bryum argenteum

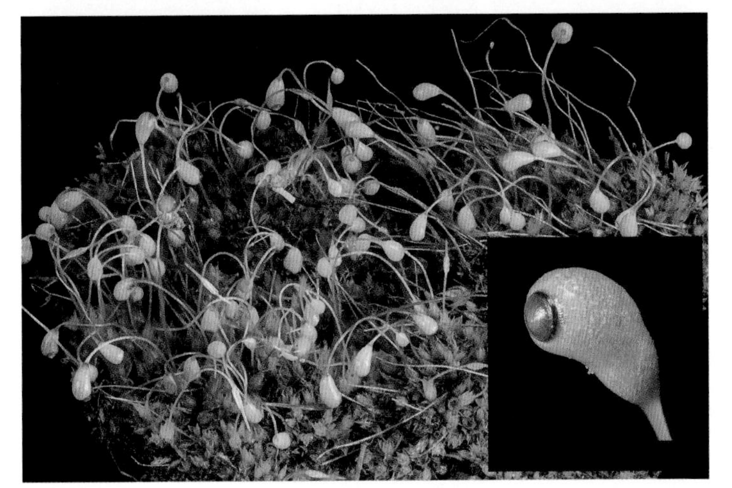

Funaria hygrometrica

Pogonatum pensilvanicum

Aphanorrhegma serratum

Pleuridium subulatum

Weissia controversa

Barbula unguiculata

Dicranella heteromalla

Ceratodon purpureus cushions & capsules

Trematodon ambiguus

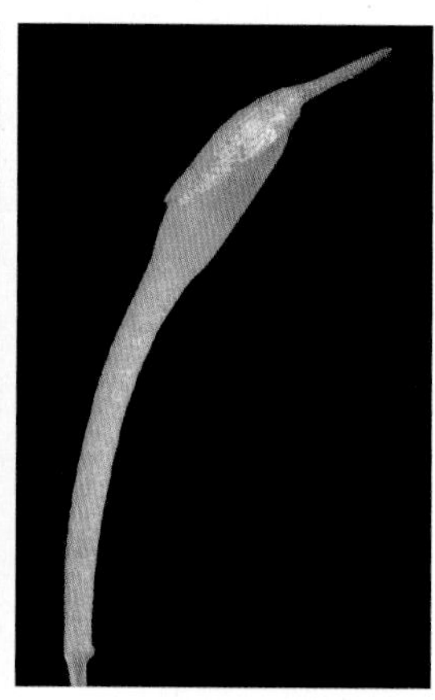

Trematodon longicollis

Polytrichum piliferum

Pogonatum dentatum

Fissidens bushii

Pseudotaxiphyllum distichaceum

Brachythecium velutinum

Campylium chrysophyllum

LIVERWORTS

Blasia pusilla

Calypogeia muelleriana

LOOKING FOR GOLD - GOBLIN'S GOLD

When you find yourself looking deep under a ledge where the light is dim, or way in a deep dark recess under a root, or even in the dim light of an old stone foundation under a barn, that's the time to be looking for gold - Goblin's Gold.

Reaching in, you grab a handful of the shiny stuff, but when you open your hand there's nothing there! You might just have found *Schistostega pennata*, a tiny delicate moss that can grow on soil in the darkest of holes. The shiny stuff you see is the persistent luminous protonema from which it arises. Sometimes you might find yourself holding a fragile, tiny, pale-green, fern-like moss, which is the sterile form and most often seen. It shrivels quite quickly when removed from the dark, moist soil where you found it. If you are lucky enough to find capsules, you'll see that they are ovoid, erect and have no peristome, or teeth, around the mouth.

LIMY SOIL

(SAW)

Limy soil can be found in a variety of places. It may be soil where groundwater has seeped through limy rocks or soil that covers some type of limy bedrock. Rich glades or limy knolls or cobbles are an example of places you might find these mosses. Generally these are dry, wooded, rounded hills with loose fragments of rocks and also some small ledges. There isn't any water flow from above, therefore the soils are often dry and thin. Grasses and sedges may be abundant, and the canopy is often thin so ample sunlight can get through. This type of place might have large moss mats of *Rhytidium rugosum, Thuidium recognitum, Tortella tortuosa, Abietinella abietina, Anomodon attenuatus* and *A. rostratus* covering the ground, thin soil over ledges and ledges themselves. On the smaller loose stones you will often find *Amblystegium varium, Homomallium adnatum* and *Platydictya confervoides.*

Another type of habitat is much wetter, having a claylike soil with a lot of drainage from above and seepage through calcareous soil or ledges. On the ground here you can find such mosses as *Timmia megapolitana, Taxiphyllum deplanatum, Ctenidium molluscum, Fissidens taxifolius, Bryoandersonia illecebra, Bryhnia graminicolor* and *Eurhynchium hians.*

Cushions of *Tortella tortuosa* covering ground between rocks

A sampling of mosses on a limy cobble

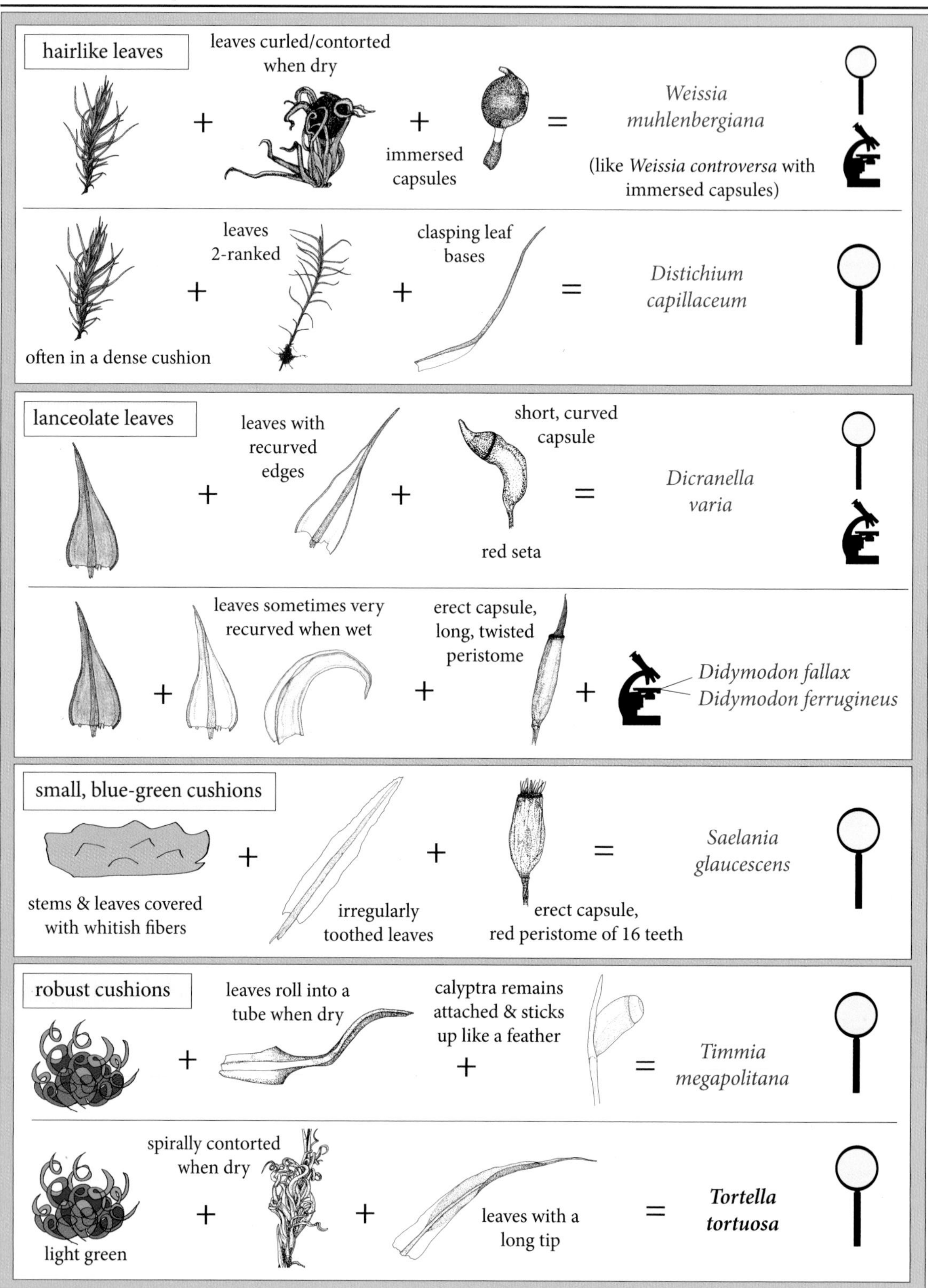
hairlike leaves
leaves curled/contorted when dry
+
+
immersed capsules
=
Weissia muhlenbergiana
(like Weissia controversa with immersed capsules)
often in a dense cushion
leaves 2-ranked
+
clasping leaf bases
+
=
Distichium capillaceum
lanceolate leaves
leaves with recurved edges
+
short, curved capsule
+
red seta
=
Dicranella varia
+
leaves sometimes very recurved when wet
erect capsule, long, twisted peristome
+
+
Didymodon fallax
Didymodon ferrugineus
small, blue-green cushions
+
+
=
Saelania glaucescens
stems & leaves covered with whitish fibers
irregularly toothed leaves
erect capsule, red peristome of 16 teeth
robust cushions
leaves roll into a tube when dry
+
calyptra remains attached & sticks up like a feather
+
=
Timmia megapolitana
spirally contorted when dry
+
+
leaves with a long tip
=
Tortella tortuosa
light green

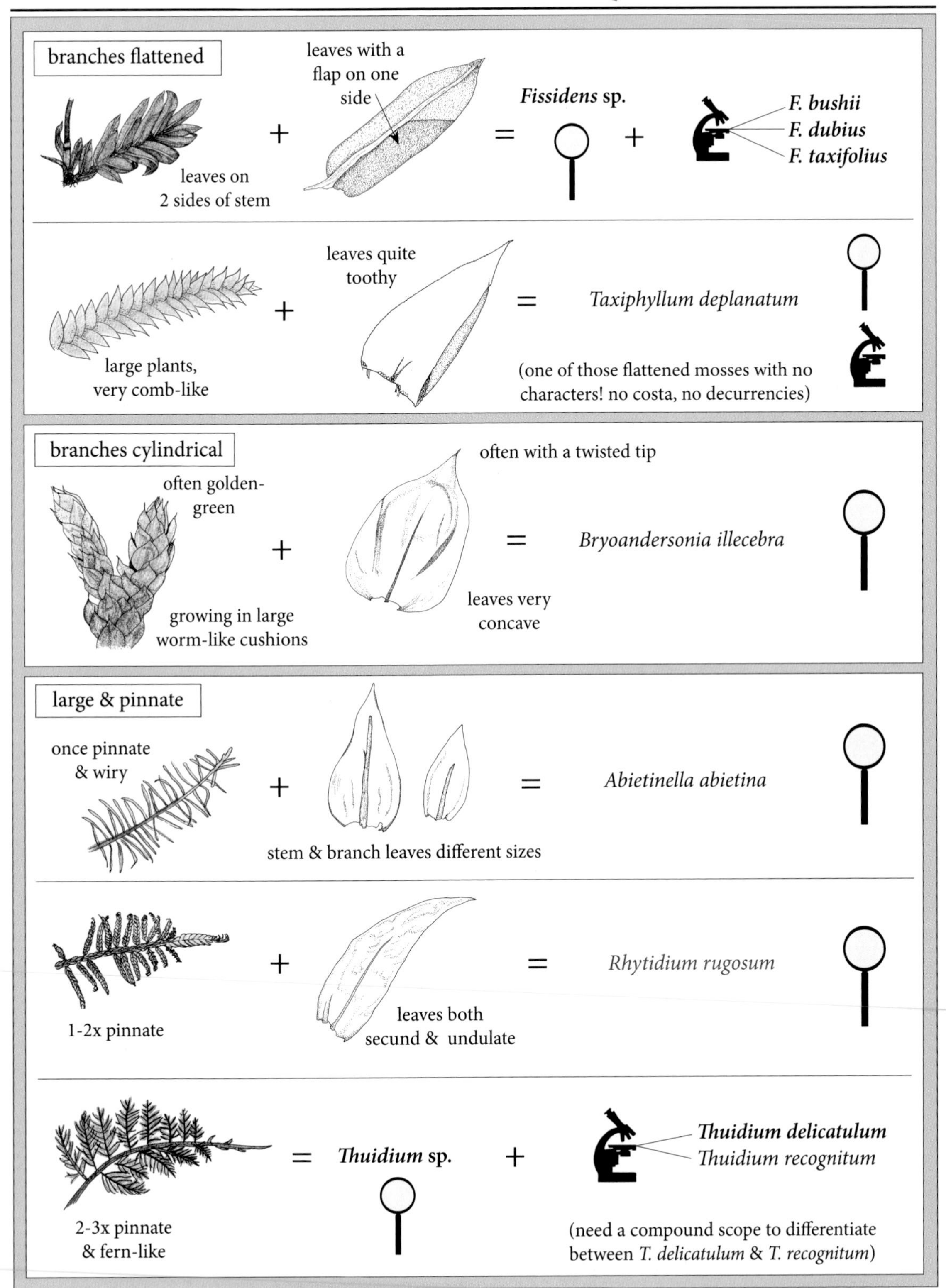
branches flattened
leaves on
2 sides of stem
+
leaves with a
flap on one
side
=
Fissidens sp.
+
F. bushii
F. dubius
F. taxifolius
large plants,
very comb-like
+
leaves quite
toothy
=
Taxiphyllum deplanatum
(one of those flattened mosses with no
characters! no costa, no decurrencies)
branches cylindrical
often golden-
green
growing in large
worm-like cushions
+
often with a twisted tip
leaves very
concave
=
Bryoandersonia illecebra
large & pinnate
once pinnate
& wiry
+
stem & branch leaves different sizes
=
Abietinella abietina
1-2x pinnate
+
leaves both
secund & undulate
=
Rhytidium rugosum
2-3x pinnate
& fern-like
=
Thuidium sp.
+
Thuidium delicatulum
Thuidium recognitum
(need a compound scope to differentiate
between T. delicatulum & T. recognitum)

Limy Soil - Quick & Dirty Field Key

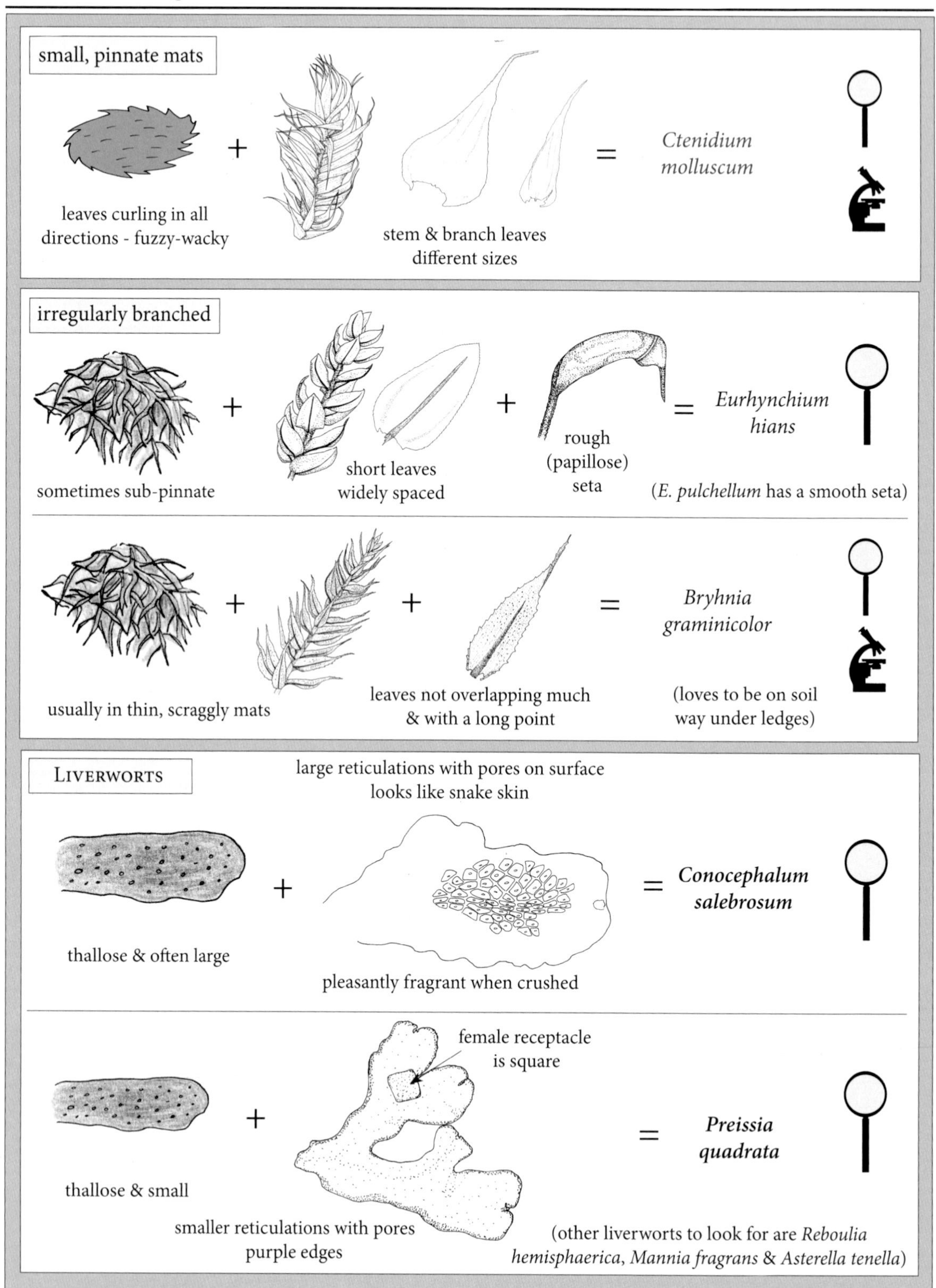

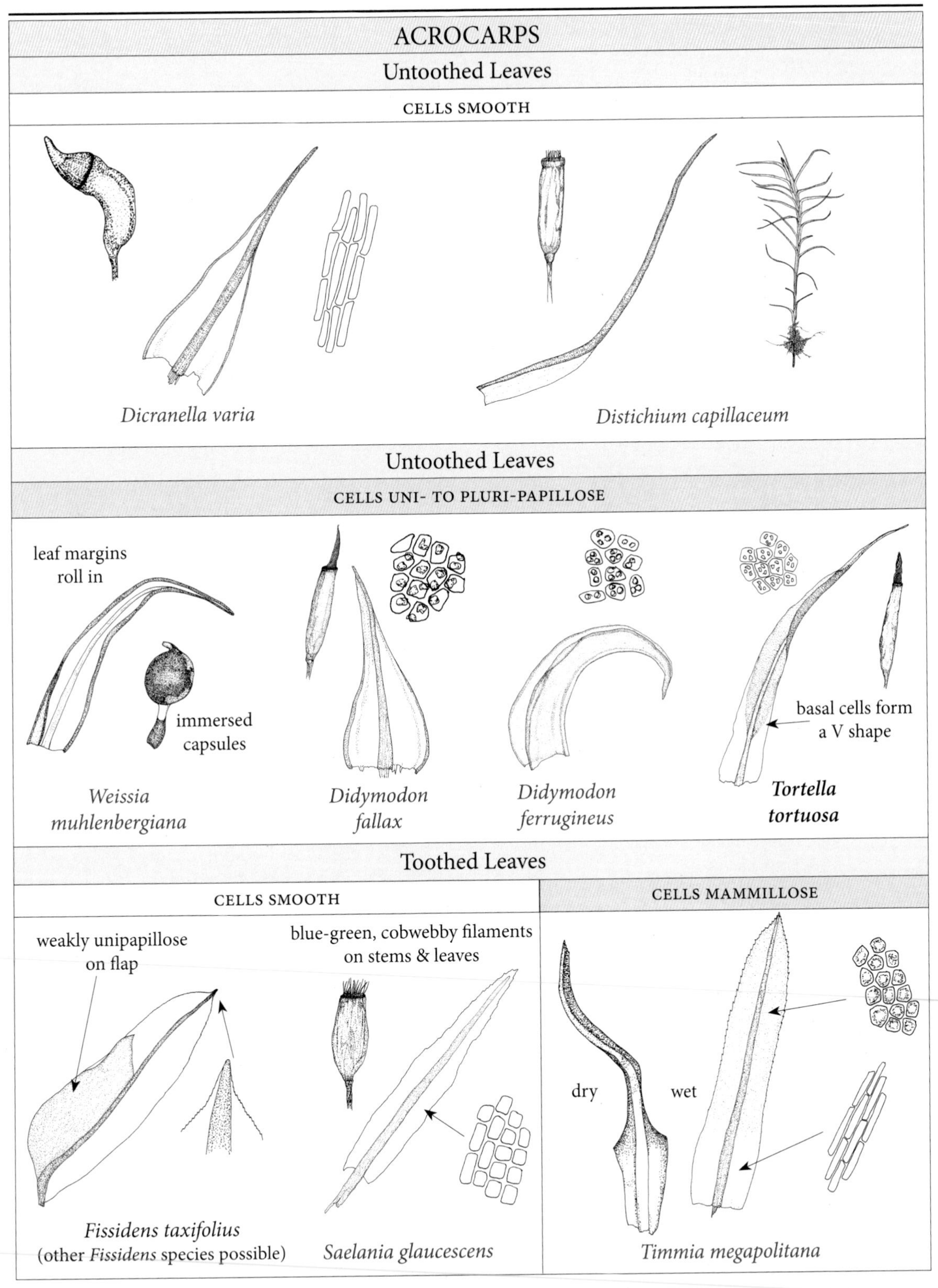
ACROCARPS
Untoothed Leaves
CELLS SMOOTH
Dicranella varia
Distichium capillaceum
Untoothed Leaves
CELLS UNI- TO PLURI-PAPILLOSE
leaf margins
roll in
immersed
capsules
Weissia
muhlenbergiana
Didymodon
fallax
Didymodon
ferrugineus
basal cells form
a V shape
Tortella
tortuosa
Toothed Leaves
CELLS SMOOTH
weakly unipapillose
on flap
blue-green, cobwebby filaments
on stems & leaves
Fissidens taxifolius
(other Fissidens species possible)
Saelania glaucescens
CELLS MAMMILLOSE
dry
wet
Timmia megapolitana

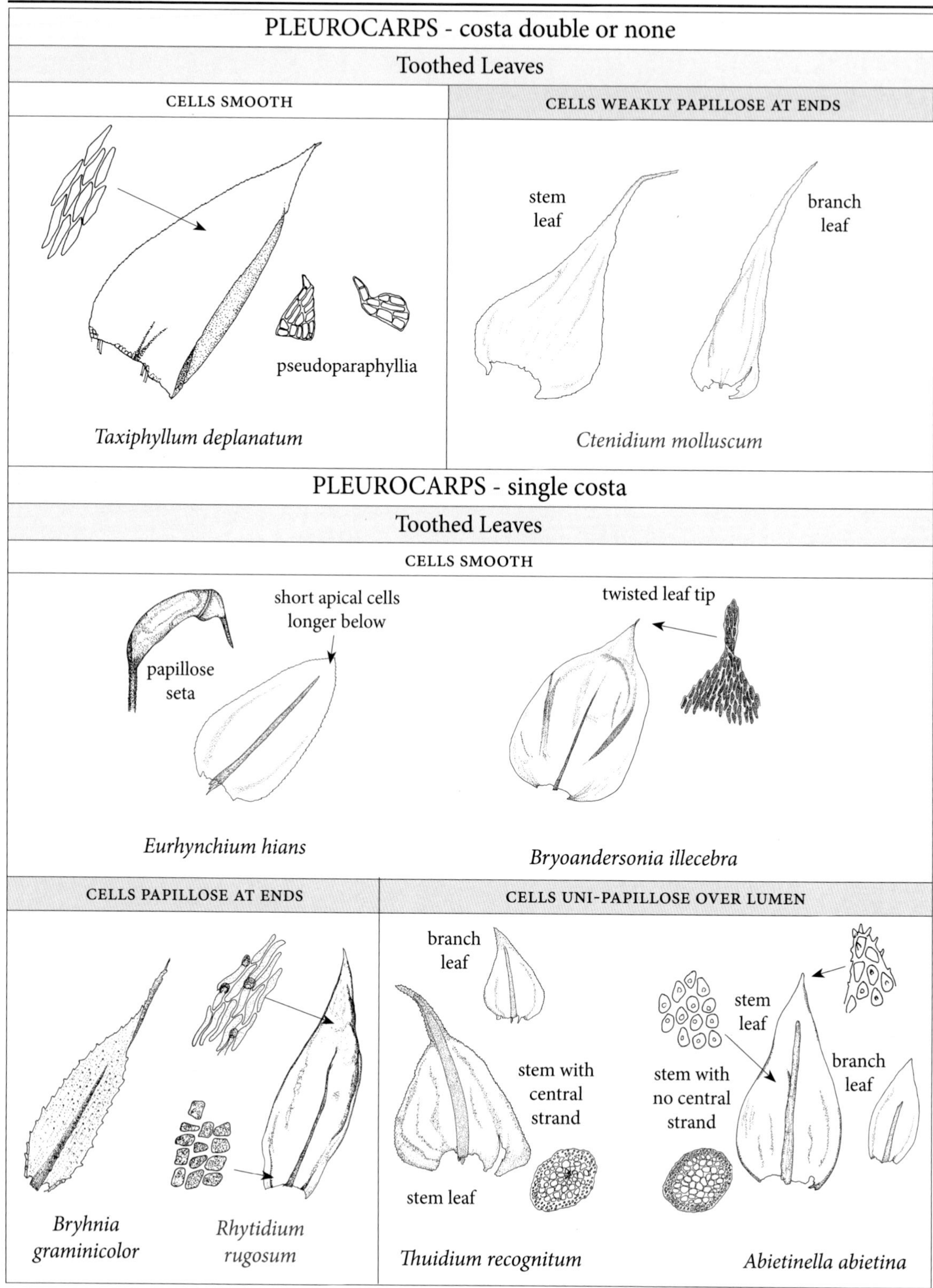
PLEUROCARPS - costa double or none
Toothed Leaves
CELLS SMOOTH
CELLS WEAKLY PAPILLOSE AT ENDS
pseudoparaphyllia
Taxiphyllum deplanatum
stem leaf
branch leaf
Ctenidium molluscum
PLEUROCARPS - single costa
Toothed Leaves
CELLS SMOOTH
short apical cells longer below
papillose seta
Eurhynchium hians
twisted leaf tip
Bryoandersonia illecebra
CELLS PAPILLOSE AT ENDS
Bryhnia graminicolor
Rhytidium rugosum
CELLS UNI-PAPILLOSE OVER LUMEN
branch leaf
stem with central strand
stem leaf
Thuidium recognitum
stem leaf
branch leaf
stem with no central strand
Abietinella abietina

Didymodon ferrugineus

Distichium capillaceum

Saelania glaucescens

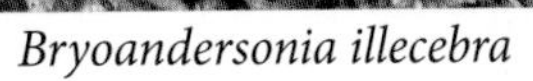

Bryoandersonia illecebra

Timmia megapolitana

Abietinella abietina

Thuidium recognitum

Ctenidium molluscum

Rhytidium rugosum

Taxiphyllum deplanatum

Conocephalum salebrosum

DRY ROCKS & LEDGES

(SAW)

Dry rocks & ledges in this section include non-calcareous boulders & ledges both in the shade and exposed. These rocks will have basically no seepage or runoff other than rainfall. Some boulders have high moss coverage, and some have uneven coverage depending on type of rock, exposure and dirtiness of the rock. Boulders can collect litter on their flat or concave surfaces, which traps water, and can form quite thick soils on the surface where mosses that prefer the forest floor can now grow, such as *Pleurozium schreberi* or *Polytrichum pallidisetum*. The bases of boulders can have quite shaded recesses that are moister than the vertical sides, and a different community of bryophytes might be found there. On the shaded underside of boulders in deciduous woods is the place to find the unusual *Campylostelium saxicola*.

The most common species found on dry, shaded boulders and ledges are *Dicranum fulvum* and *Plagiothecium* species.

Dicranum fulvum grows in dark green to blackish-green curly cushions, and *Plagiothecium* forms flattened waterfalls.

Ledges often have at least some kind of seepage or surface flow, but the ones covered in this section will have minimal. Soil can also form in crevices and steps of the ledges where water seeps out. Channels and grooves of runoff will provide a wetter location for many bryophytes, and usually a ledge has a greater diversity of species due to the varied moisture levels. Like boulders, the bases or lowest parts of ledges are also moister and can have different species than the sides or top. Shaded cracks and crevices are great places to look for such species as *Homalia trichomanoides* or *Thamnobryum alleghaniense*.

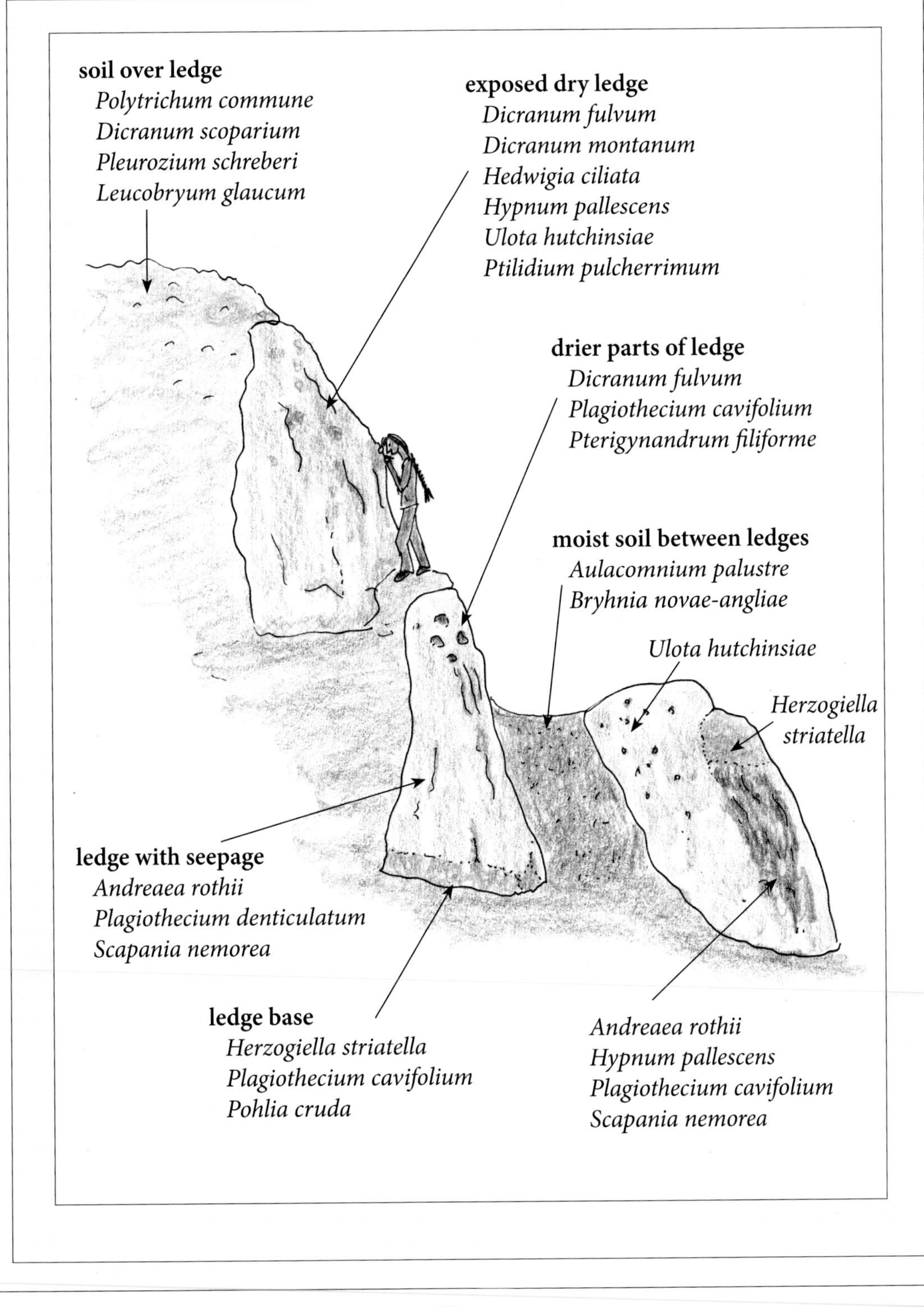
soil over ledge
Polytrichum commune
Dicranum scoparium
Pleurozium schreberi
Leucobryum glaucum
exposed dry ledge
Dicranum fulvum
Dicranum montanum
Hedwigia ciliata
Hypnum pallescens
Ulota hutchinsiae
Ptilidium pulcherrimum
drier parts of ledge
Dicranum fulvum
Plagiothecium cavifolium
Pterigynandrum filiforme
moist soil between ledges
Aulacomnium palustre
Bryhnia novae-angliae
Ulota hutchinsiae
Herzogiella striatella
ledge with seepage
Andreaea rothii
Plagiothecium denticulatum
Scapania nemorea
ledge base
Herzogiella striatella
Plagiothecium cavifolium
Pohlia cruda
Andreaea rothii
Hypnum pallescens
Plagiothecium cavifolium
Scapania nemorea

Dry Rocks & Ledges - Quick & Dirty Field Key

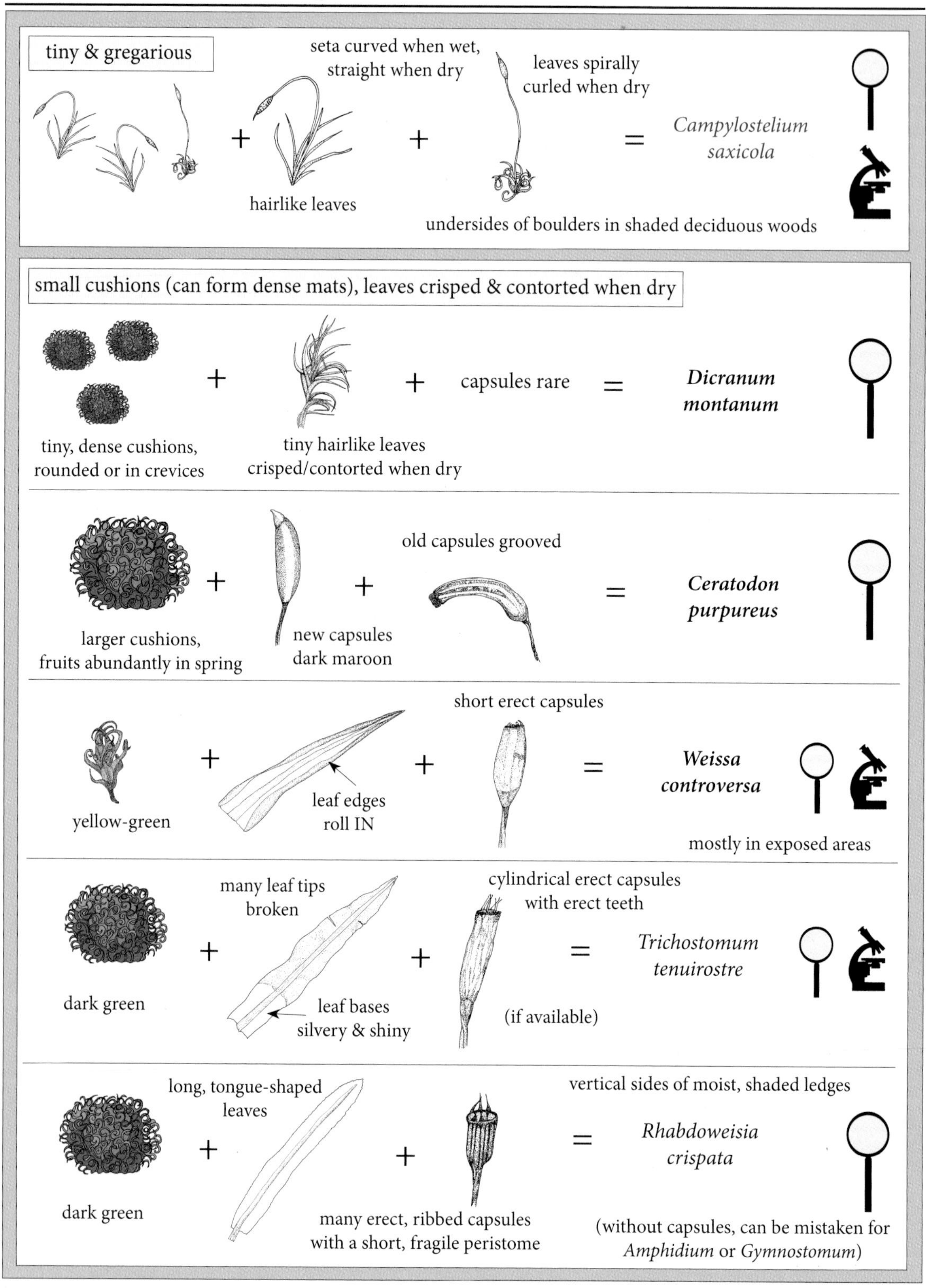

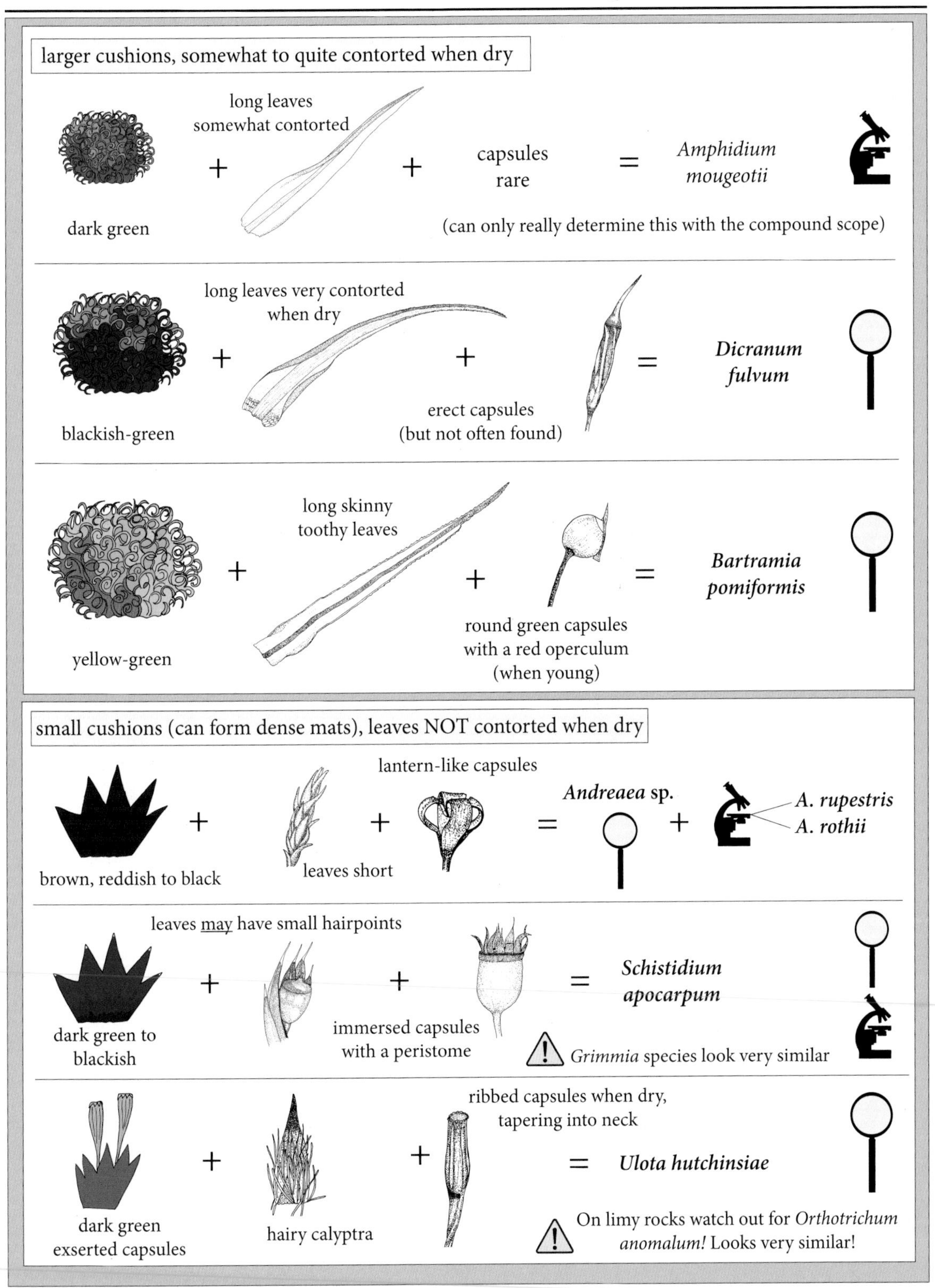
larger cushions, somewhat to quite contorted when dry
dark green
long leaves somewhat contorted
capsules rare
Amphidium mougeotii
(can only really determine this with the compound scope)
blackish-green
long leaves very contorted when dry
erect capsules (but not often found)
Dicranum fulvum
yellow-green
long skinny toothy leaves
round green capsules with a red operculum (when young)
Bartramia pomiformis
small cushions (can form dense mats), leaves NOT contorted when dry
brown, reddish to black
leaves short
lantern-like capsules
Andreaea sp.
A. rupestris
A. rothii
dark green to blackish
leaves may have small hairpoints
immersed capsules with a peristome
Schistidium apocarpum
Grimmia species look very similar
dark green exserted capsules
hairy calyptra
ribbed capsules when dry, tapering into neck
Ulota hutchinsiae
On limy rocks watch out for Orthotrichum anomalum! Looks very similar!

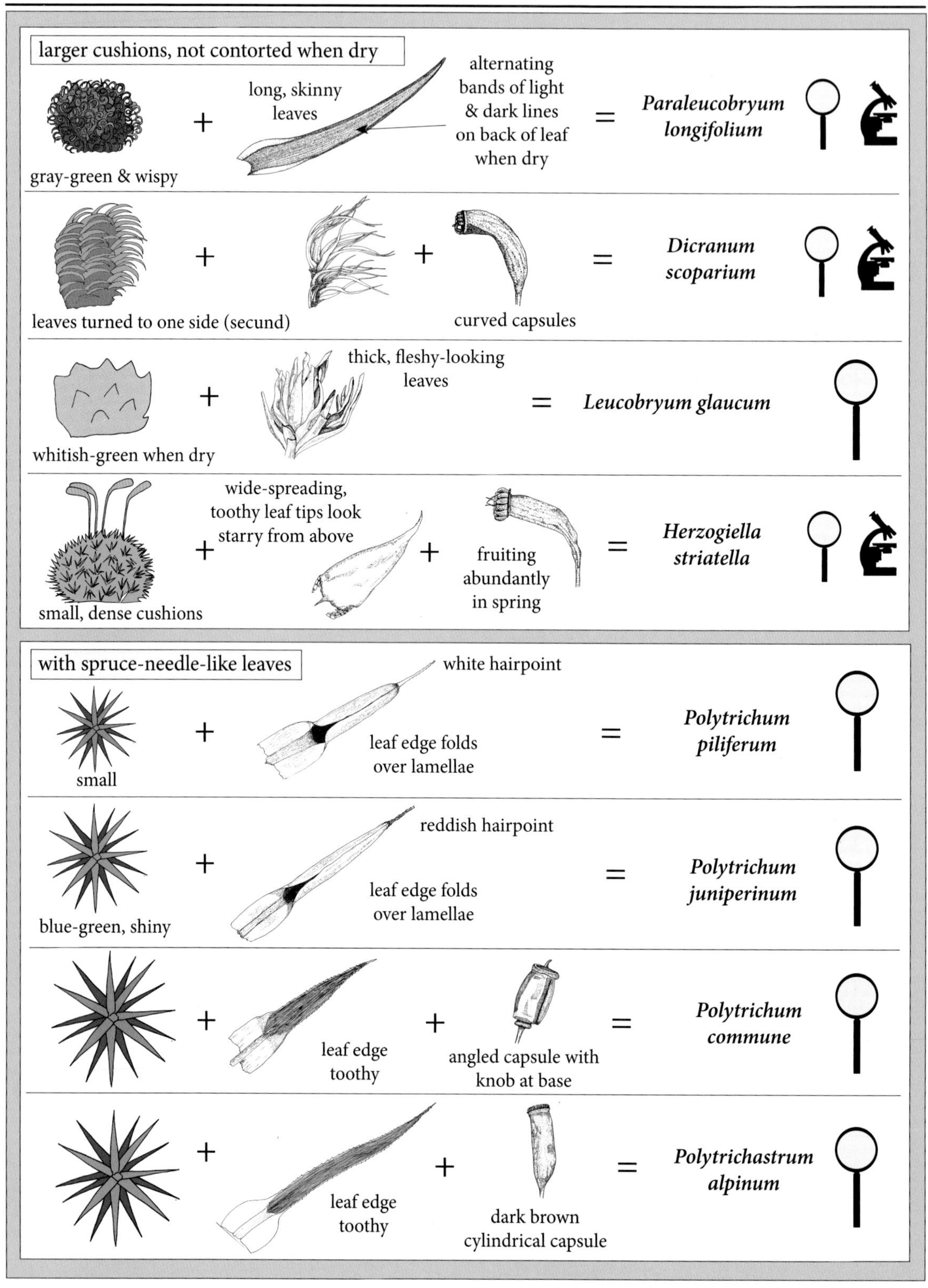
larger cushions, not contorted when dry
gray-green & wispy
long, skinny leaves
alternating bands of light & dark lines on back of leaf when dry
Paraleucobryum longifolium
leaves turned to one side (secund)
curved capsules
Dicranum scoparium
whitish-green when dry
thick, fleshy-looking leaves
Leucobryum glaucum
small, dense cushions
wide-spreading, toothy leaf tips look starry from above
fruiting abundantly in spring
Herzogiella striatella
with spruce-needle-like leaves
small
white hairpoint
leaf edge folds over lamellae
Polytrichum piliferum
blue-green, shiny
reddish hairpoint
leaf edge folds over lamellae
Polytrichum juniperinum
leaf edge toothy
angled capsule with knob at base
Polytrichum commune
leaf edge toothy
dark brown cylindrical capsule
Polytrichastrum alpinum

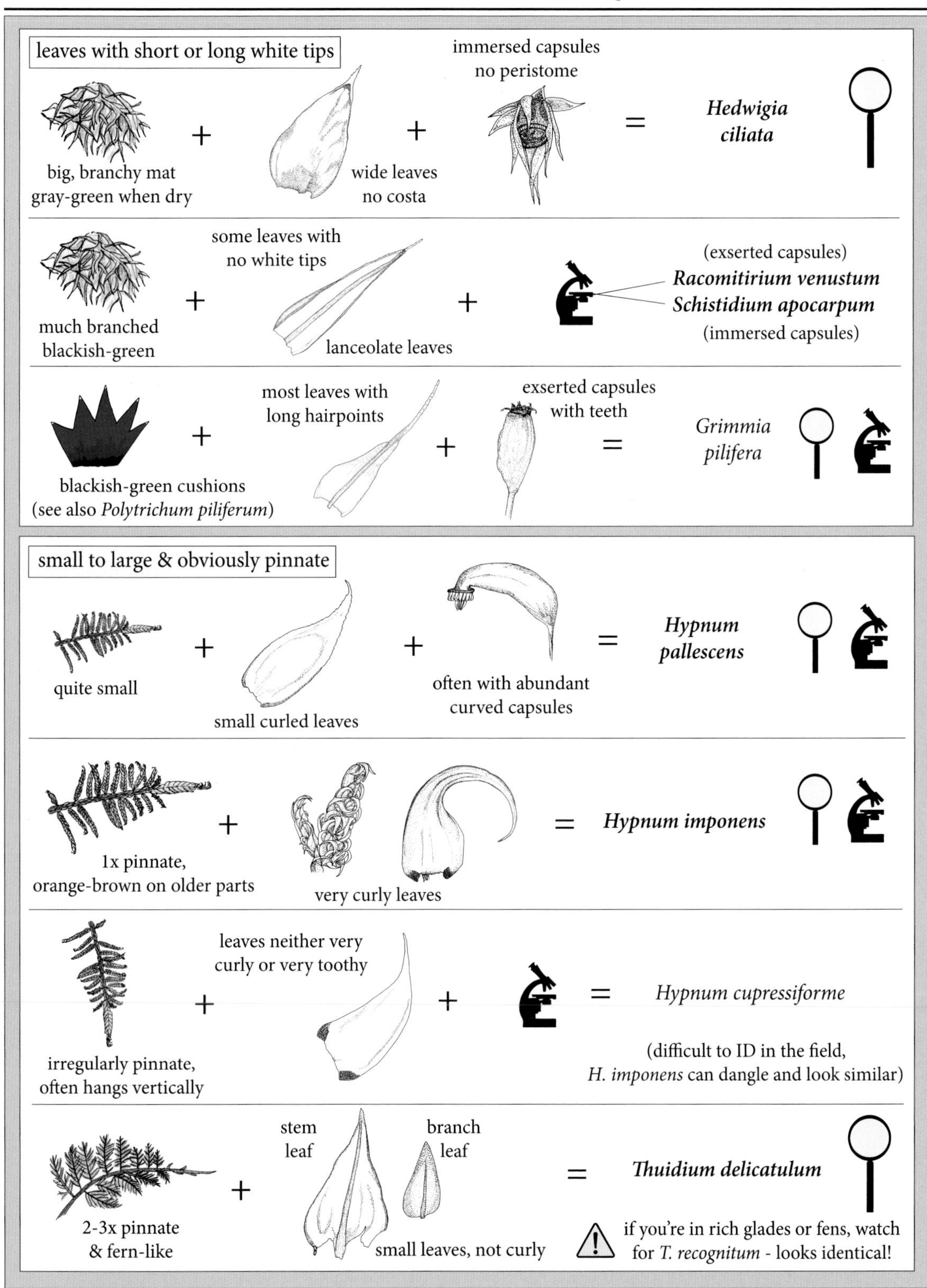
leaves with short or long white tips
big, branchy mat
gray-green when dry
+
wide leaves
no costa
+
immersed capsules
no peristome
=
Hedwigia
ciliata
much branched
blackish-green
+
some leaves with
no white tips
lanceolate leaves
+
(exserted capsules)
Racomitirium venustum
Schistidium apocarpum
(immersed capsules)
blackish-green cushions
(see also Polytrichum piliferum)
+
most leaves with
long hairpoints
+
exserted capsules
with teeth
=
Grimmia
pilifera
small to large & obviously pinnate
quite small
+
small curled leaves
+
often with abundant
curved capsules
=
Hypnum
pallescens
1x pinnate,
orange-brown on older parts
+
very curly leaves
=
Hypnum imponens
irregularly pinnate,
often hangs vertically
+
leaves neither very
curly or very toothy
+
=
Hypnum cupressiforme
(difficult to ID in the field,
H. imponens can dangle and look similar)
2-3x pinnate
& fern-like
+
stem
leaf
branch
leaf
small leaves, not curly
=
Thuidium delicatulum
if you're in rich glades or fens, watch
for T. recognitum - looks identical!

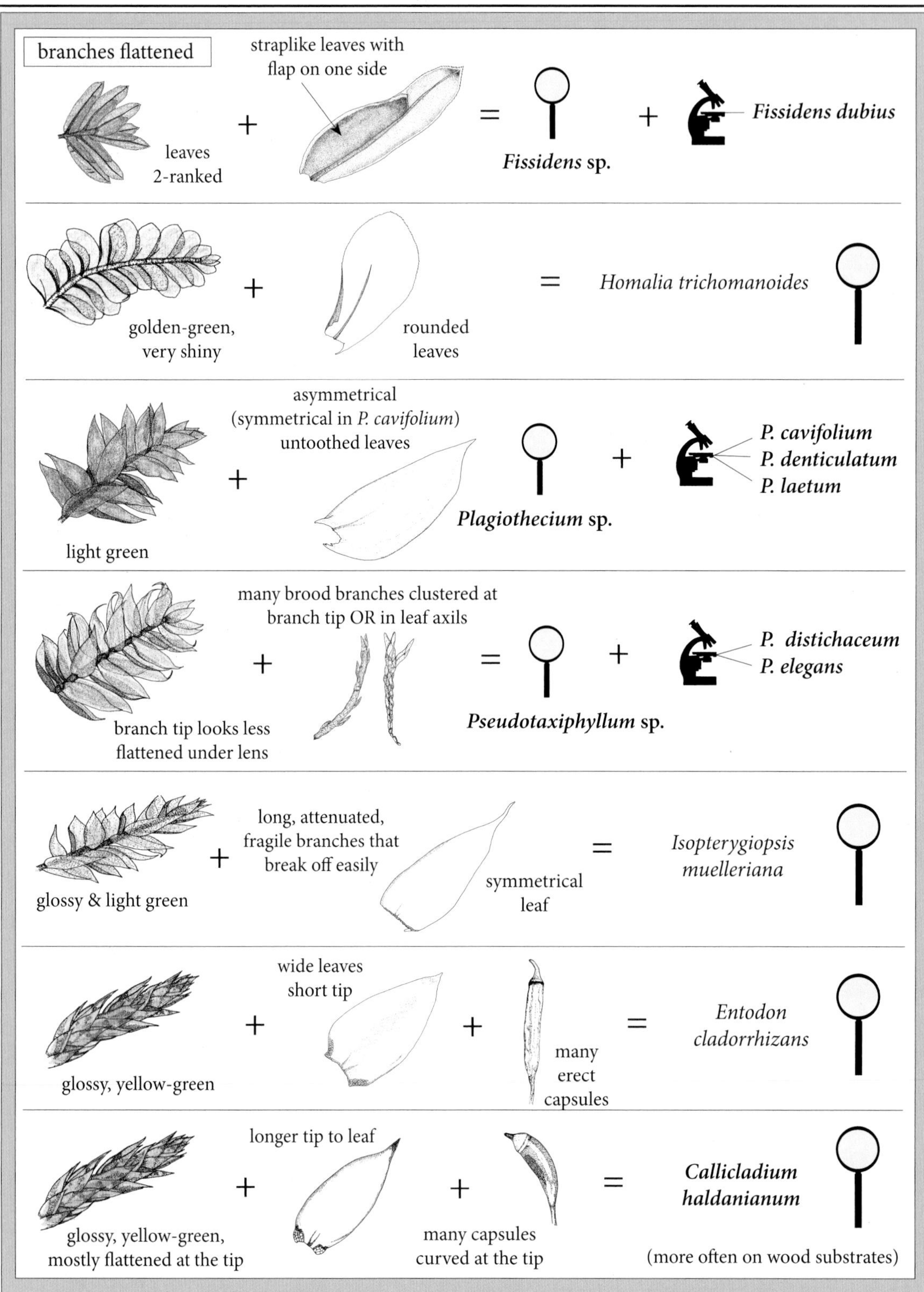
branches flattened
leaves 2-ranked
straplike leaves with flap on one side
Fissidens sp.
Fissidens dubius
golden-green, very shiny
rounded leaves
Homalia trichomanoides
light green
asymmetrical (symmetrical in P. cavifolium) untoothed leaves
Plagiothecium sp.
P. cavifolium
P. denticulatum
P. laetum
branch tip looks less flattened under lens
many brood branches clustered at branch tip OR in leaf axils
Pseudotaxiphyllum sp.
P. distichaceum
P. elegans
glossy & light green
long, attenuated, fragile branches that break off easily
symmetrical leaf
Isopterygiopsis muelleriana
glossy, yellow-green
wide leaves short tip
many erect capsules
Entodon cladorrhizans
glossy, yellow-green, mostly flattened at the tip
longer tip to leaf
many capsules curved at the tip
Callicladium haldanianum
(more often on wood substrates)

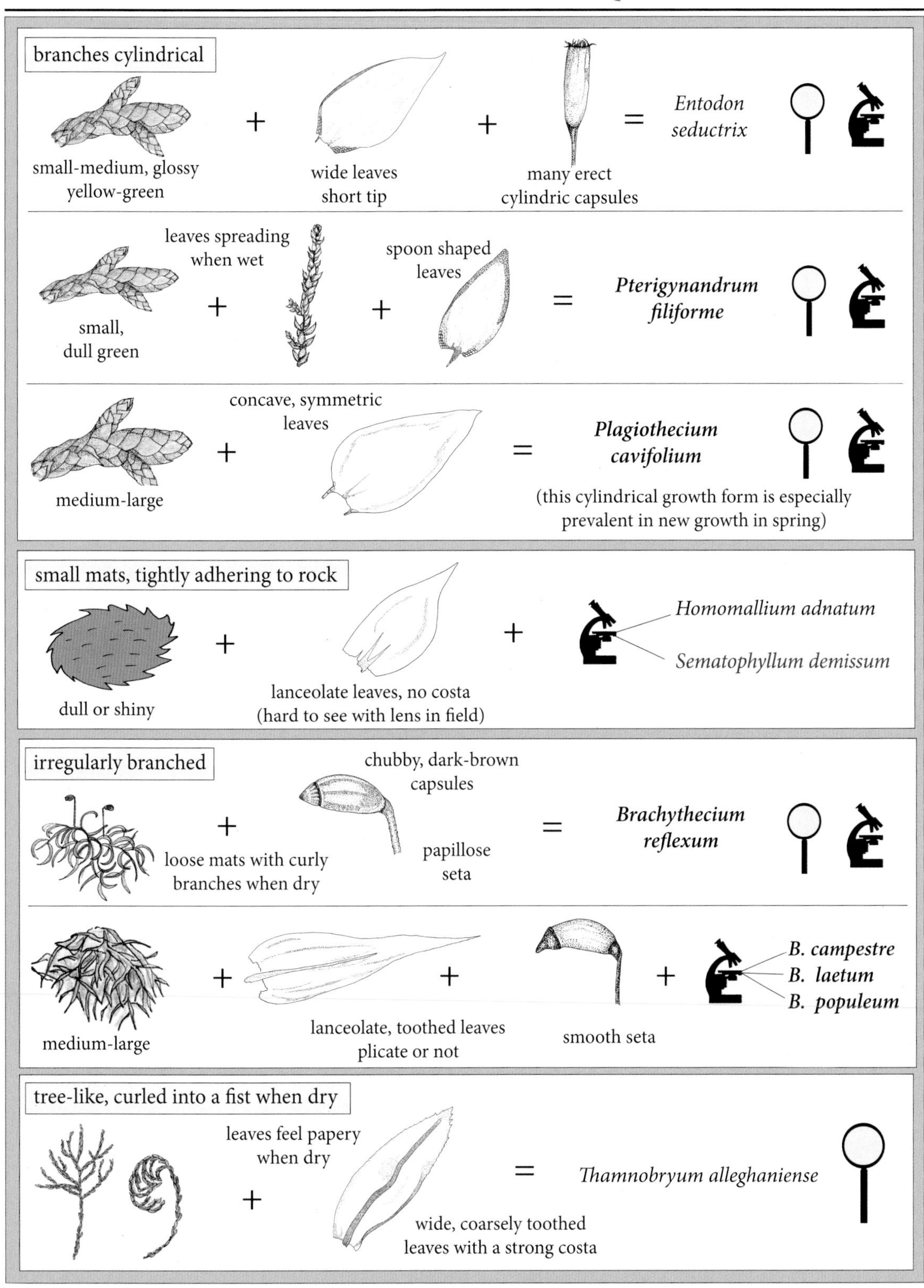

branches cylindrical
small-medium, glossy yellow-green
+
wide leaves short tip
+
many erect cylindric capsules
=
Entodon seductrix
small, dull green
leaves spreading when wet
+
spoon shaped leaves
+
=
Pterigynandrum filiforme
medium-large
+
concave, symmetric leaves
=
Plagiothecium cavifolium
(this cylindrical growth form is especially prevalent in new growth in spring)
small mats, tightly adhering to rock
dull or shiny
+
lanceolate leaves, no costa (hard to see with lens in field)
+
Homomallium adnatum
Sematophyllum demissum
irregularly branched
+
loose mats with curly branches when dry
chubby, dark-brown capsules
papillose seta
=
Brachythecium reflexum
medium-large
+
lanceolate, toothed leaves plicate or not
+
smooth seta
+
B. campestre
B. laetum
B. populeum
tree-like, curled into a fist when dry
leaves feel papery when dry
+
wide, coarsely toothed leaves with a strong costa
=
Thamnobryum alleghaniense

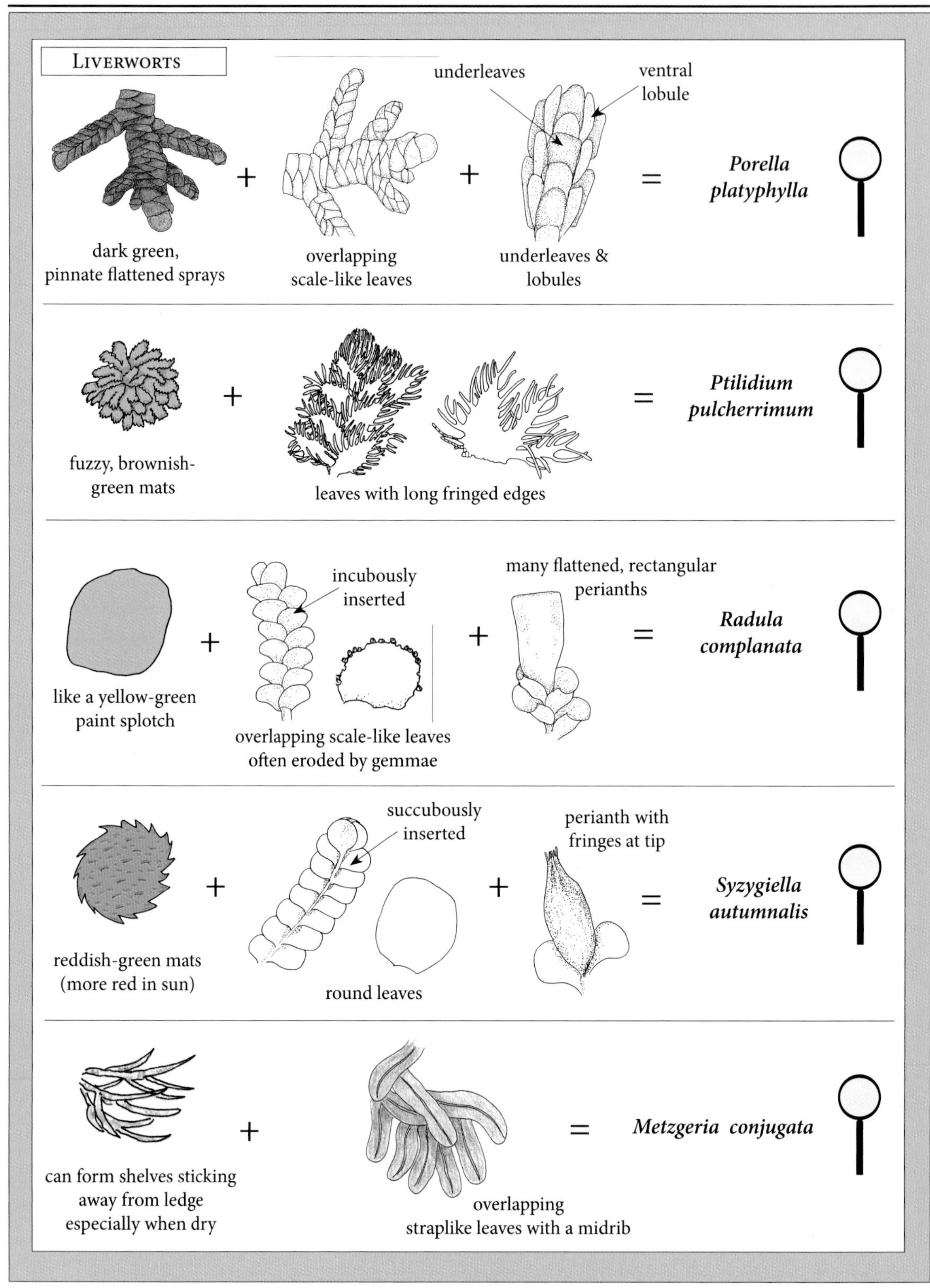
Liverworts
underleaves
ventral lobule
dark green, pinnate flattened sprays
+
overlapping scale-like leaves
+
underleaves & lobules
=
Porella platyphylla
fuzzy, brownish-green mats
+
leaves with long fringed edges
=
Ptilidium pulcherrimum
like a yellow-green paint splotch
+
incubously inserted
overlapping scale-like leaves often eroded by gemmae
+
many flattened, rectangular perianths
=
Radula complanata
reddish-green mats (more red in sun)
+
succubously inserted
round leaves
+
perianth with fringes at tip
=
Syzygiella autumnalis
can form shelves sticking away from ledge especially when dry
+
overlapping straplike leaves with a midrib
=
Metzgeria conjugata

ACROCARPS

Untoothed Leaves

CELLS SMOOTH

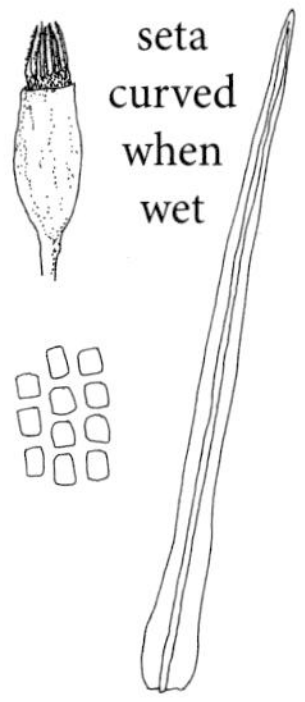

seta curved when wet

Campylostelium saxicola

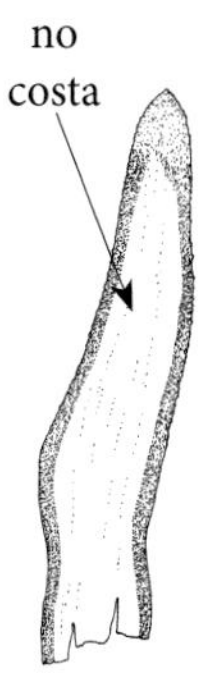

no costa

Andreaea rupestris

leaves sometimes roughened at back

Andreaea rothii

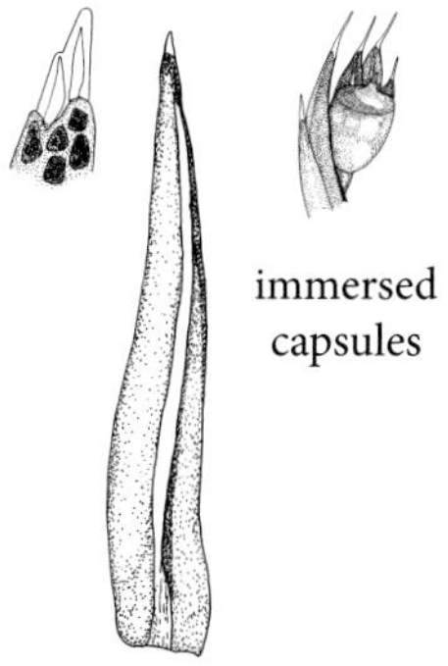

immersed capsules

Schistidium apocarpum

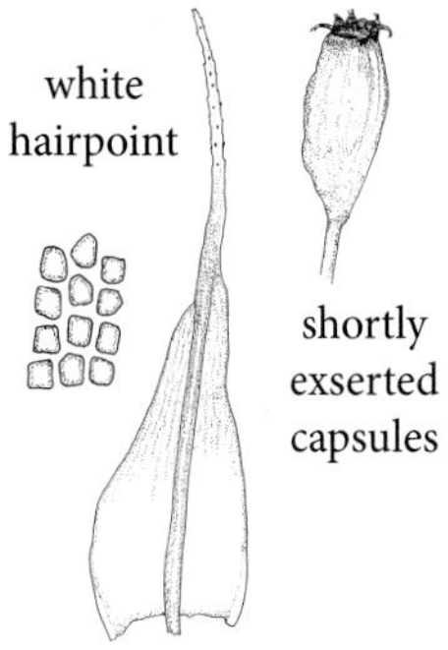

white hairpoint

shortly exserted capsules

Grimmia pilifera

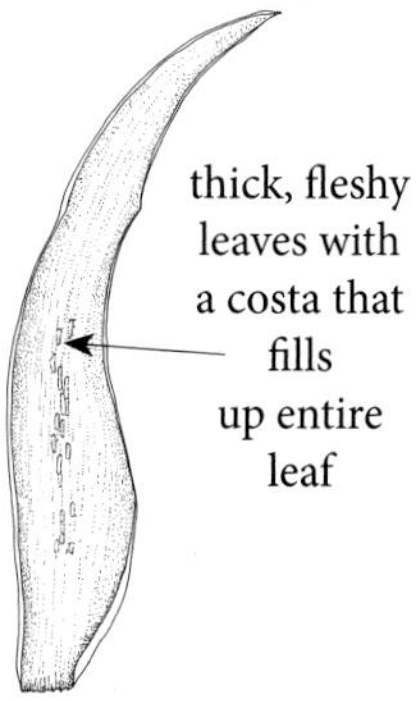

thick, fleshy leaves with a costa that fills up entire leaf

Leucobryum glaucum

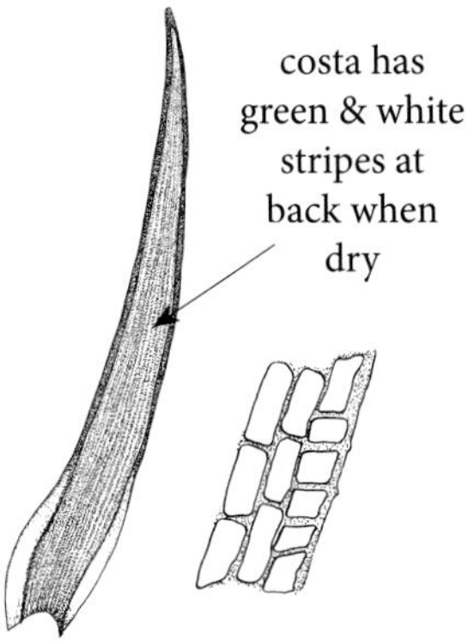

costa has green & white stripes at back when dry

Paraleucobryum longifolium

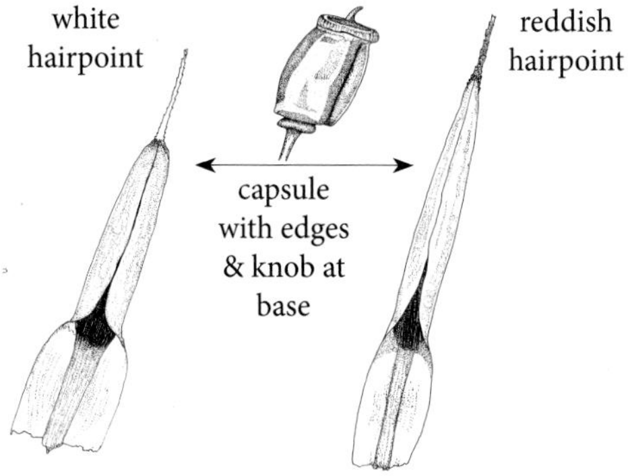

white hairpoint

capsule with edges & knob at base

Polytrichum piliferum

reddish hairpoint

Polytrichum juniperinum

CELLS UNI-PAPILLOSE

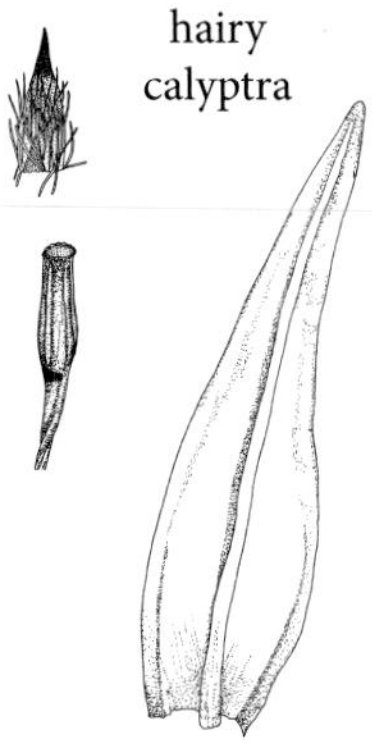

hairy calyptra

Ulota hutchinsiae

CELLS PLURI-PAPILLOSE

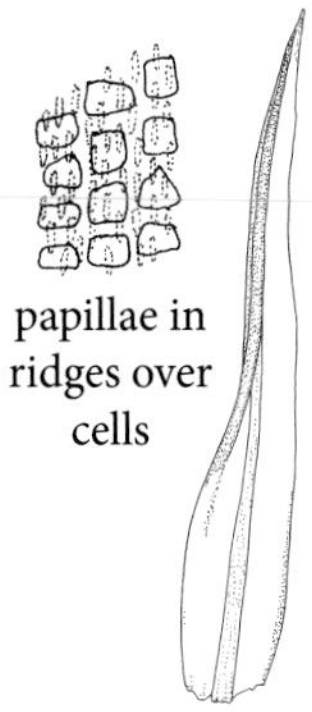

papillae in ridges over cells

Amphidium mougeotii

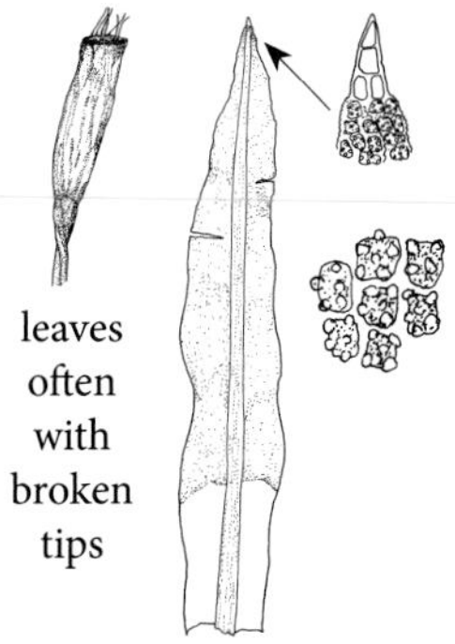

leaves often with broken tips

Trichostomum tenuirostre

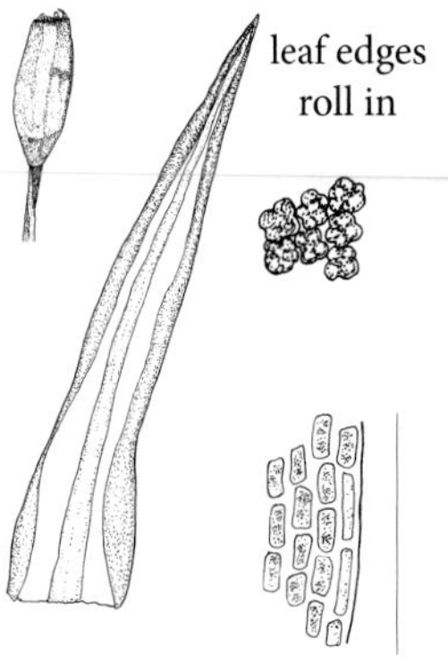

leaf edges roll in

Weissia controversa

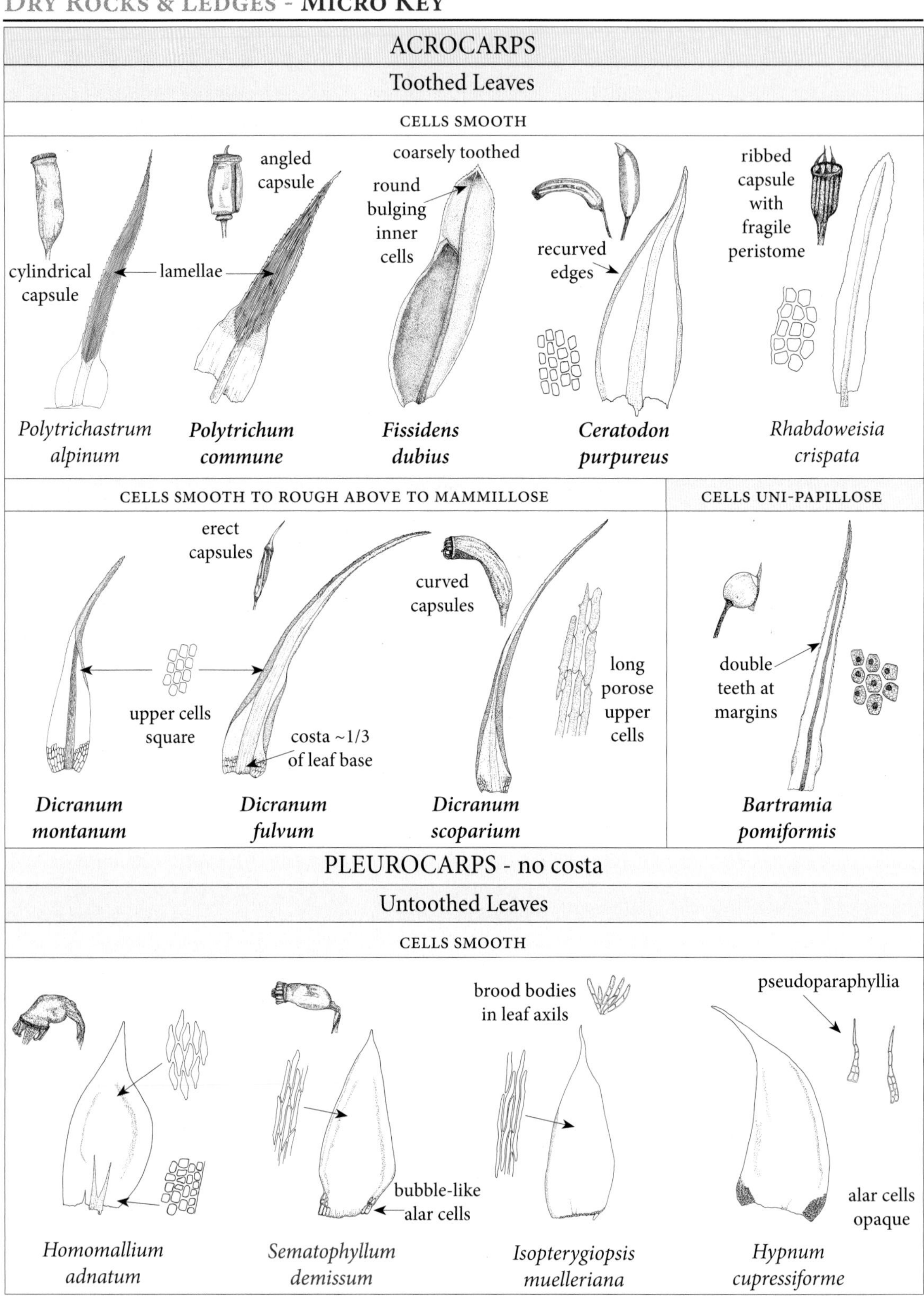
ACROCARPS
Toothed Leaves
CELLS SMOOTH
cylindrical capsule
lamellae
angled capsule
coarsely toothed
round bulging inner cells
recurved edges
ribbed capsule with fragile peristome
Polytrichastrum alpinum
Polytrichum commune
Fissidens dubius
Ceratodon purpureus
Rhabdoweisia crispata
CELLS SMOOTH TO ROUGH ABOVE TO MAMMILLOSE
CELLS UNI-PAPILLOSE
erect capsules
curved capsules
upper cells square
costa ~1/3 of leaf base
long porose upper cells
double teeth at margins
Dicranum montanum
Dicranum fulvum
Dicranum scoparium
Bartramia pomiformis
PLEUROCARPS - no costa
Untoothed Leaves
CELLS SMOOTH
brood bodies in leaf axils
pseudoparaphyllia
bubble-like alar cells
alar cells opaque
Homomallium adnatum
Sematophyllum demissum
Isopterygiopsis muelleriana
Hypnum cupressiforme

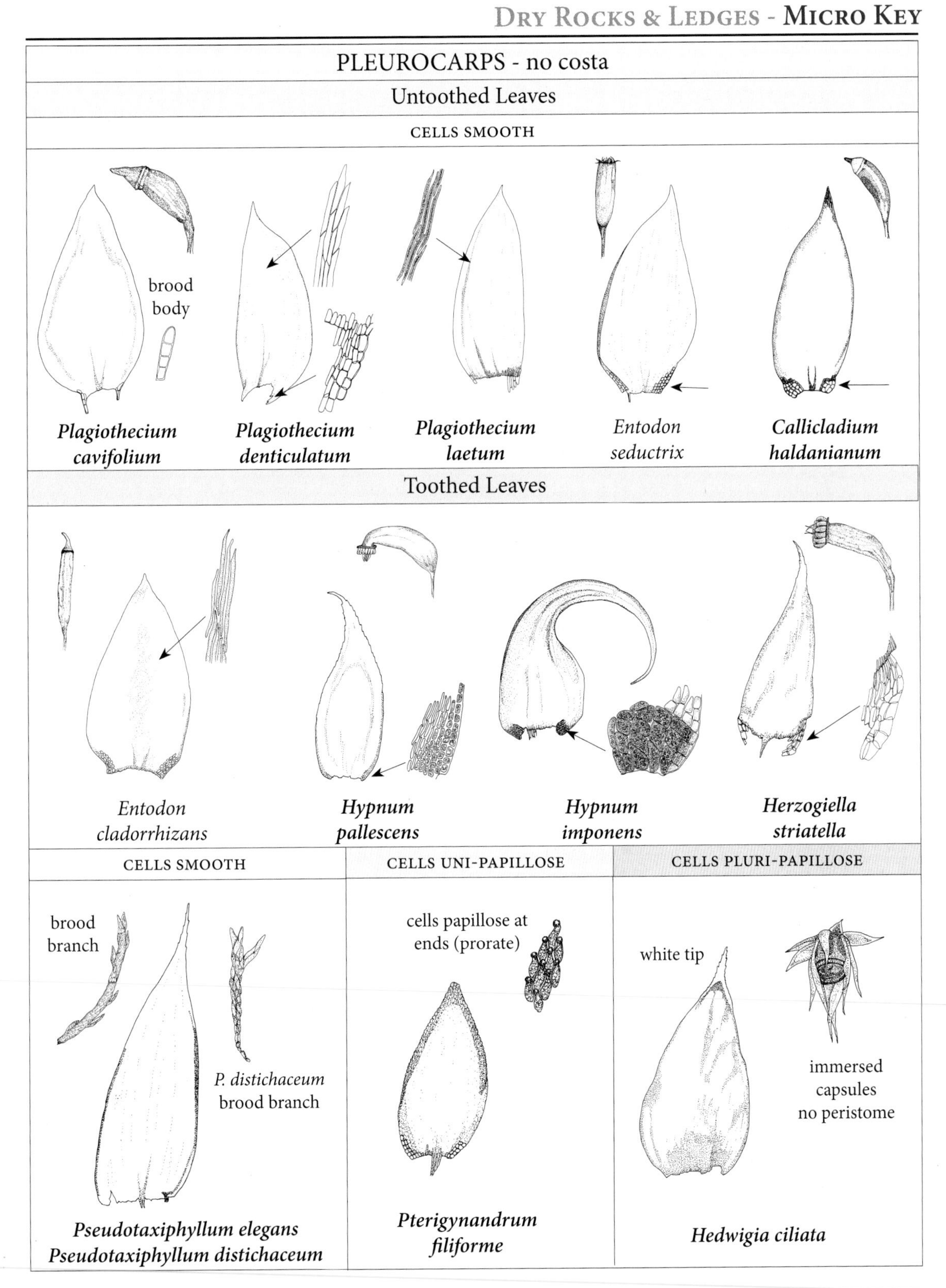
PLEUROCARPS - no costa
Untoothed Leaves
CELLS SMOOTH
brood body
Plagiothecium cavifolium
Plagiothecium denticulatum
Plagiothecium laetum
Entodon seductrix
Callicladium haldanianum
Toothed Leaves
Entodon cladorrhizans
Hypnum pallescens
Hypnum imponens
Herzogiella striatella
CELLS SMOOTH
CELLS UNI-PAPILLOSE
CELLS PLURI-PAPILLOSE
brood branch
P. distichaceum brood branch
Pseudotaxiphyllum elegans
Pseudotaxiphyllum distichaceum
cells papillose at ends (prorate)
Pterigynandrum filiforme
white tip
immersed capsules no peristome
Hedwigia ciliata

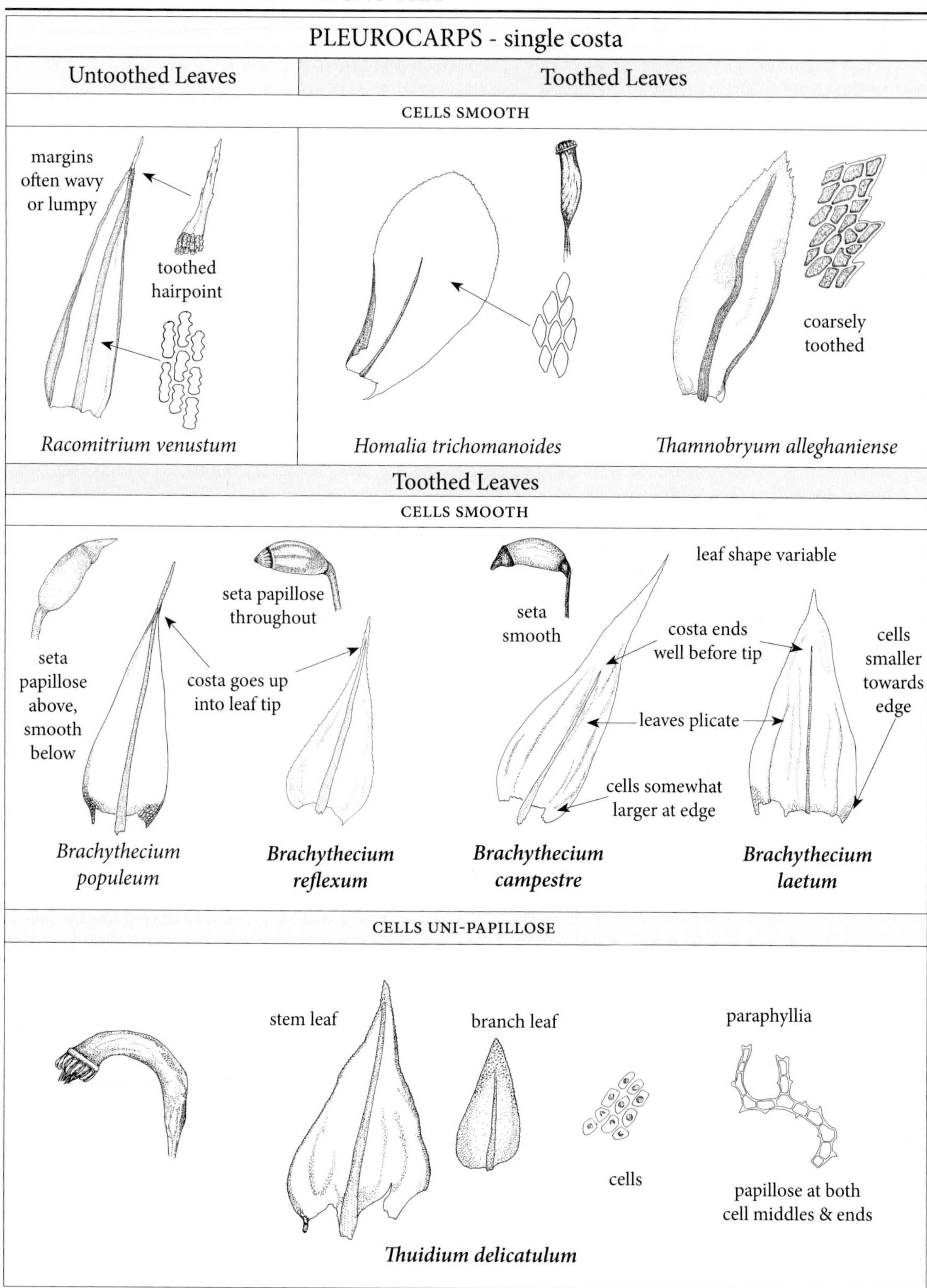
PLEUROCARPS - single costa
Untoothed Leaves
Toothed Leaves
CELLS SMOOTH
margins often wavy or lumpy
toothed hairpoint
Racomitrium venustum
Homalia trichomanoides
coarsely toothed
Thamnobryum alleghaniense
Toothed Leaves
CELLS SMOOTH
seta papillose above, smooth below
seta papillose throughout
costa goes up into leaf tip
Brachythecium populeum
Brachythecium reflexum
seta smooth
leaf shape variable
costa ends well before tip
leaves plicate
cells somewhat larger at edge
cells smaller towards edge
Brachythecium campestre
Brachythecium laetum
CELLS UNI-PAPILLOSE
stem leaf
branch leaf
cells
paraphyllia
papillose at both cell middles & ends
Thuidium delicatulum

Dicranum fulvum

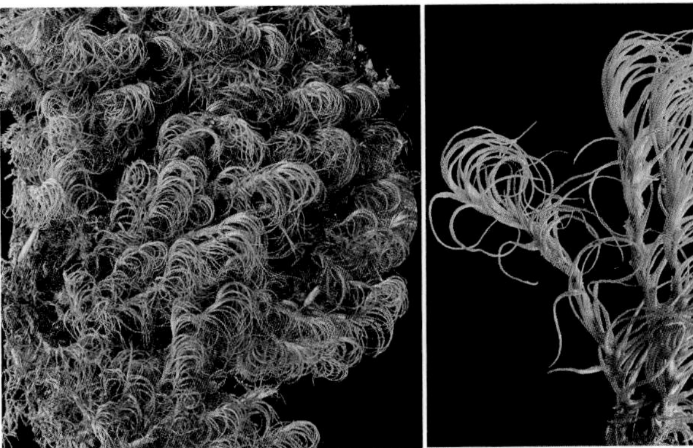

Paraleucobryum longifolium

Andreaea rupestris

Schistidium apocarpum

Ulota hutchinsiae

Bartramia pomiformis

Grimmia pilifera — ***Polytrichum piliferum***

Hedwigia ciliata

Polytrichum juniperinum

Polytrichum commune

Polytrichastrum alpinum

Hypnum pallescens *Hypnum imponens* *Thuidium delicatulum*

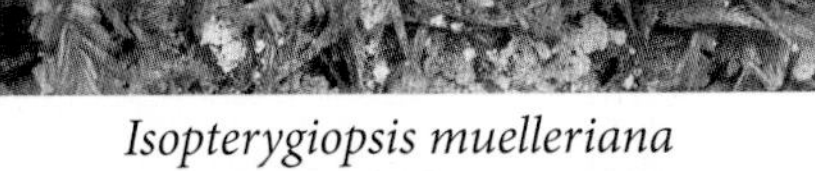

Isopterygiopsis muelleriana

Entodon cladorrhizans

Homalia trichomanoides

Plagiothecium cavifolium

Entodon seductrix

Fissidens dubius

Thamnobryum alleghaniense

Porella platyphylla

Ptilidium pulcherrimum

Syzygiella autumnalis

Metzgeria conjugata habit & close-up

LIMY ROCKS & LEDGES

(SAW)

Rocks and ledges that are high in calcium or magnesium affect the community of mosses growing there. The bryophytes in this section will be found only on this type of rock (what I call limy rocks) or have seepage through such rock, making seepage crevices a highly desirable location for these lime-loving mosses.

A dry, exposed limestone ledge might have high coverage but only with a few species, such as *Anomodon attenuatus*, *A. rostratus*, *Tortella tortuosa* and *Abietinella*. A shaded ledge that has some seepage, however, will have a much greater diversity, and it is possible to find five different species within a few inches!

A ledge may be predominantly acidic but with calcareous inclusions or seepage cracks, and therefore you will find both acid and lime-loving species. These ledges will have the highest diversity of all.

Without knowing geology, how can one tell if a rock is limy or not? By knowing one specific moss - *Anomodon attenuatus*. This large and easily identifiable moss grows on virtually any kind of limy rock, and whenever I see this species I automatically start searching for other lime-loving species.

Anomodon attenuatus - dry

Anomodon attenuatus - wet

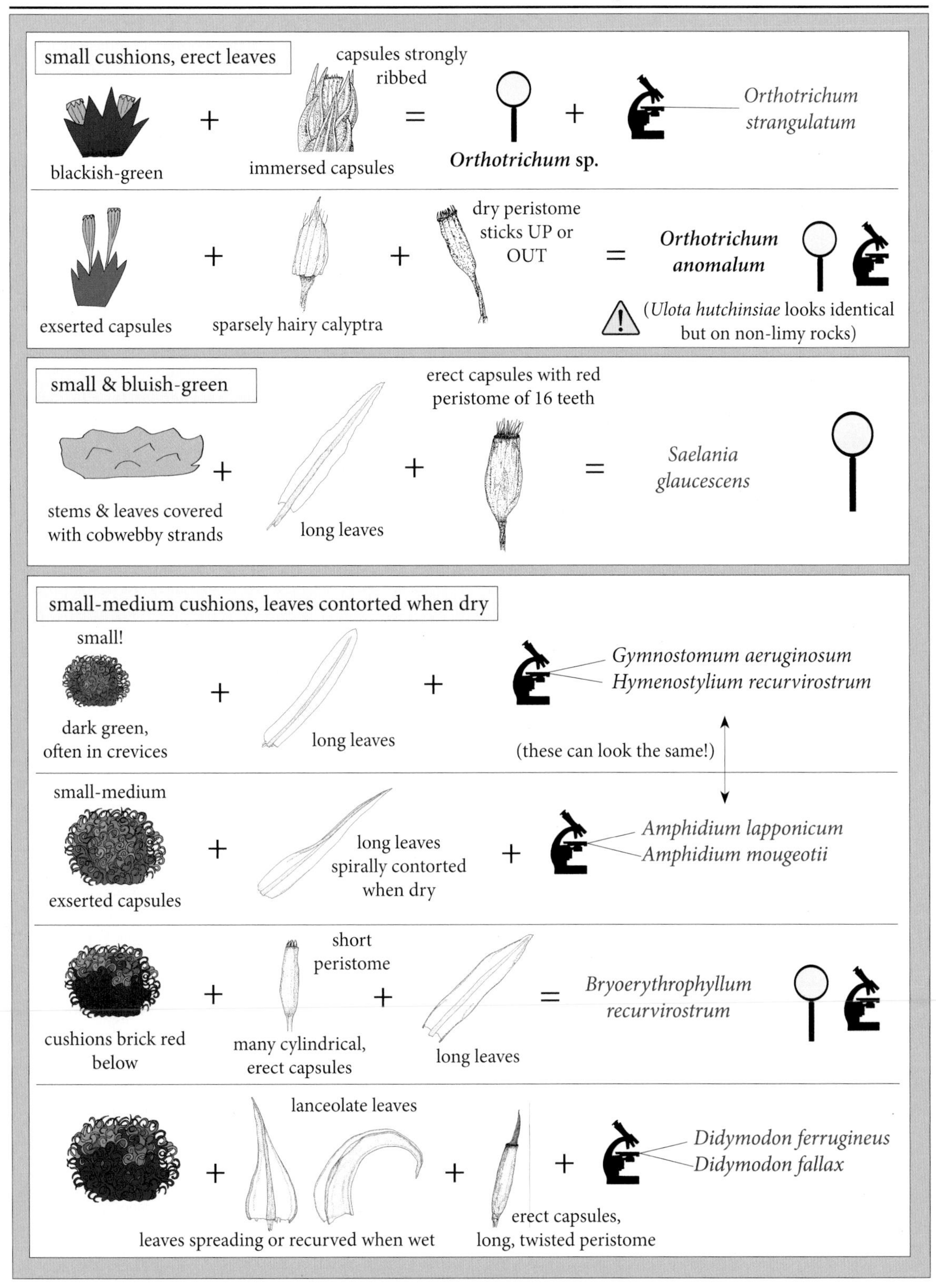
small cushions, erect leaves
blackish-green
+
capsules strongly ribbed
immersed capsules
=
Orthotrichum sp.
+
Orthotrichum strangulatum
exserted capsules
+
sparsely hairy calyptra
+
dry peristome sticks UP or OUT
=
Orthotrichum anomalum
(Ulota hutchinsiae looks identical but on non-limy rocks)
small & bluish-green
stems & leaves covered with cobwebby strands
+
long leaves
+
erect capsules with red peristome of 16 teeth
=
Saelania glaucescens
small-medium cushions, leaves contorted when dry
small!
dark green, often in crevices
+
long leaves
+
Gymnostomum aeruginosum
Hymenostylium recurvirostrum
(these can look the same!)
small-medium
exserted capsules
+
long leaves spirally contorted when dry
+
Amphidium lapponicum
Amphidium mougeotii
cushions brick red below
+
short peristome
many cylindrical, erect capsules
+
long leaves
=
Bryoerythrophyllum recurvirostrum
+
lanceolate leaves
leaves spreading or recurved when wet
+
erect capsules, long, twisted peristome
+
Didymodon ferrugineus
Didymodon fallax

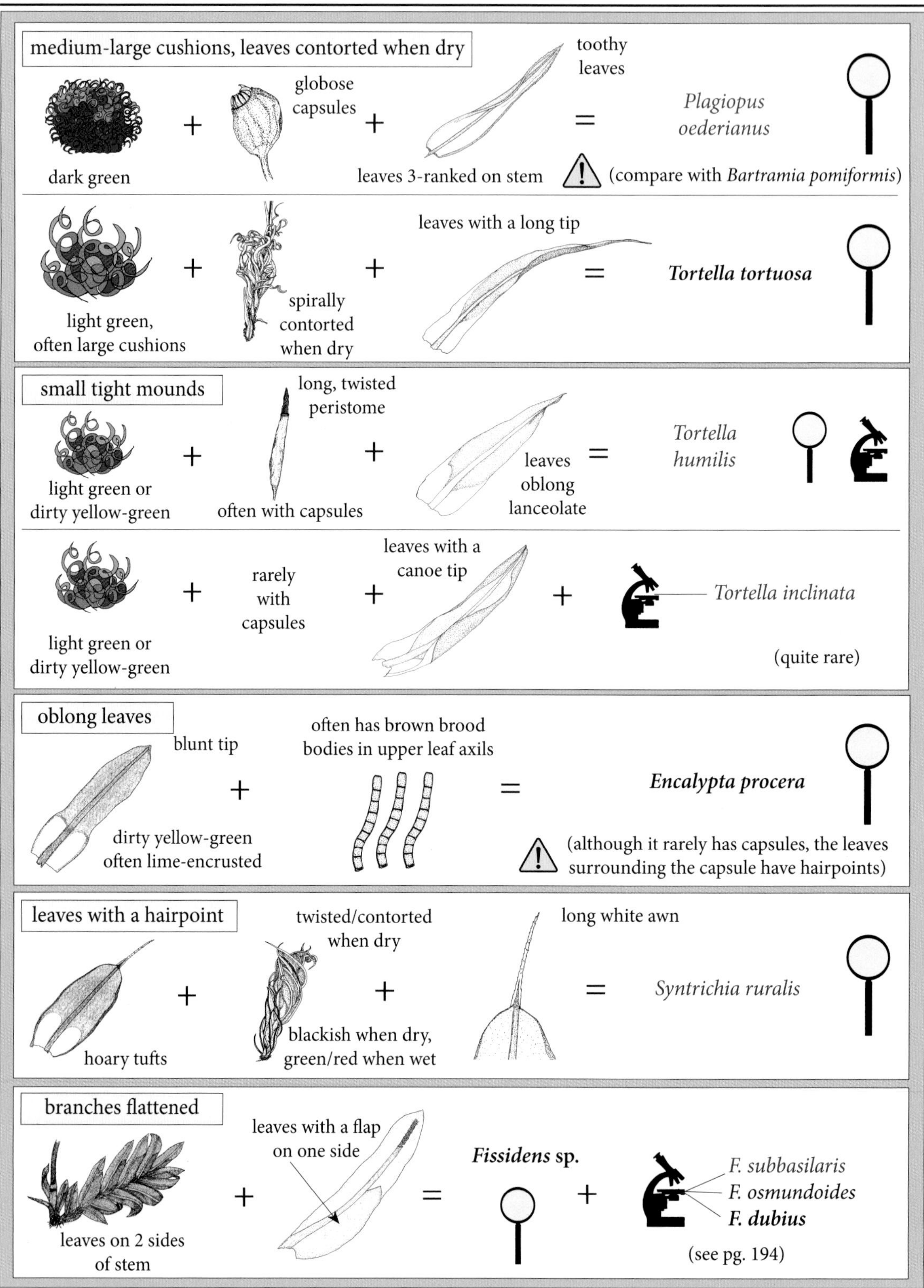
medium-large cushions, leaves contorted when dry
dark green
globose capsules
toothy leaves
leaves 3-ranked on stem
Plagiopus oederianus
(compare with Bartramia pomiformis)
light green, often large cushions
spirally contorted when dry
leaves with a long tip
Tortella tortuosa
small tight mounds
light green or dirty yellow-green
long, twisted peristome
often with capsules
leaves oblong lanceolate
Tortella humilis
light green or dirty yellow-green
rarely with capsules
leaves with a canoe tip
Tortella inclinata
(quite rare)
oblong leaves
blunt tip
dirty yellow-green often lime-encrusted
often has brown brood bodies in upper leaf axils
Encalypta procera
(although it rarely has capsules, the leaves surrounding the capsule have hairpoints)
leaves with a hairpoint
hoary tufts
twisted/contorted when dry
blackish when dry, green/red when wet
long white awn
Syntrichia ruralis
branches flattened
leaves on 2 sides of stem
leaves with a flap on one side
Fissidens sp.
F. subbasilaris
F. osmundoides
F. dubius
(see pg. 194)

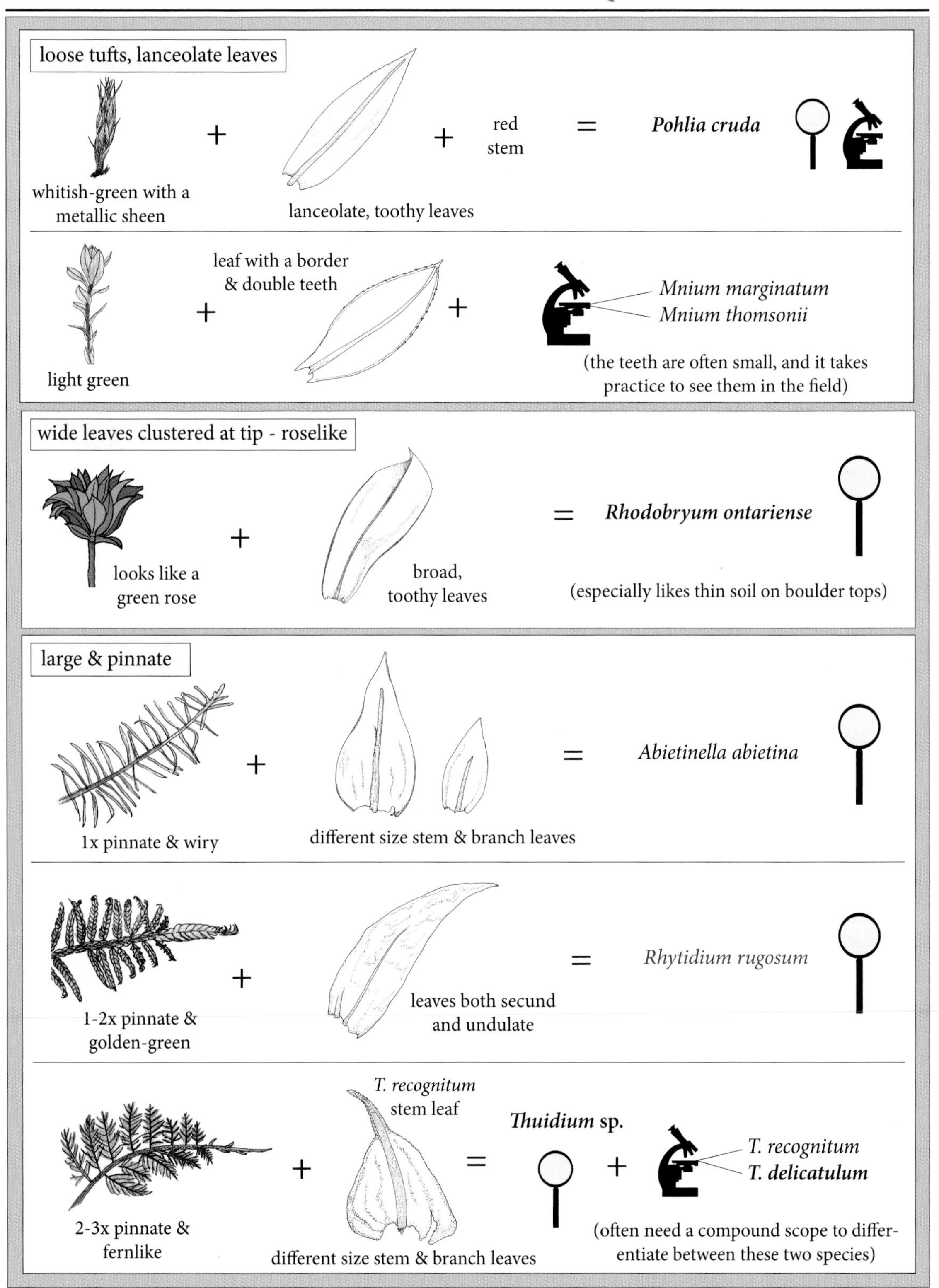
loose tufts, lanceolate leaves
whitish-green with a metallic sheen
+
lanceolate, toothy leaves
+
red stem
=
Pohlia cruda
light green
+
leaf with a border & double teeth
+
Mnium marginatum
Mnium thomsonii
(the teeth are often small, and it takes practice to see them in the field)
wide leaves clustered at tip - roselike
looks like a green rose
+
broad, toothy leaves
=
Rhodobryum ontariense
(especially likes thin soil on boulder tops)
large & pinnate
1x pinnate & wiry
+
different size stem & branch leaves
=
Abietinella abietina
1-2x pinnate & golden-green
+
leaves both secund and undulate
=
Rhytidium rugosum
2-3x pinnate & fernlike
+
T. recognitum stem leaf
different size stem & branch leaves
=
Thuidium sp.
+
T. recognitum
T. delicatulum
(often need a compound scope to differentiate between these two species)

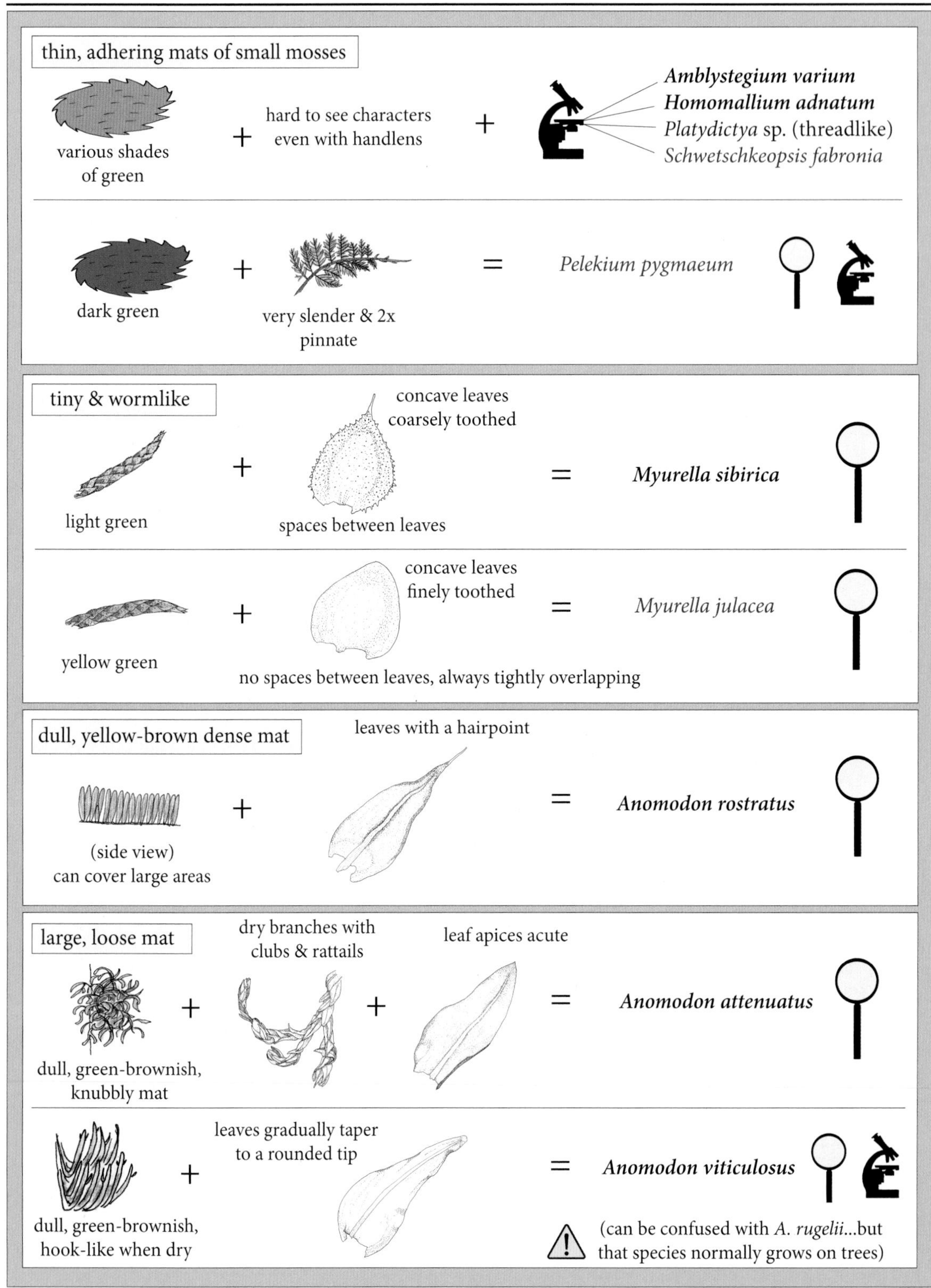
thin, adhering mats of small mosses
various shades of green
+
hard to see characters even with handlens
+
Amblystegium varium
Homomallium adnatum
Platydictya sp. (threadlike)
Schwetschkeopsis fabronia
dark green
+
very slender & 2x pinnate
=
Pelekium pygmaeum
tiny & wormlike
light green
+
concave leaves coarsely toothed
spaces between leaves
=
Myurella sibirica
yellow green
+
concave leaves finely toothed
no spaces between leaves, always tightly overlapping
=
Myurella julacea
dull, yellow-brown dense mat
(side view) can cover large areas
+
leaves with a hairpoint
=
Anomodon rostratus
large, loose mat
dull, green-brownish, knubbly mat
+
dry branches with clubs & rattails
+
leaf apices acute
=
Anomodon attenuatus
dull, green-brownish, hook-like when dry
+
leaves gradually taper to a rounded tip
=
Anomodon viticulosus
(can be confused with *A. rugelii*...but that species normally grows on trees)

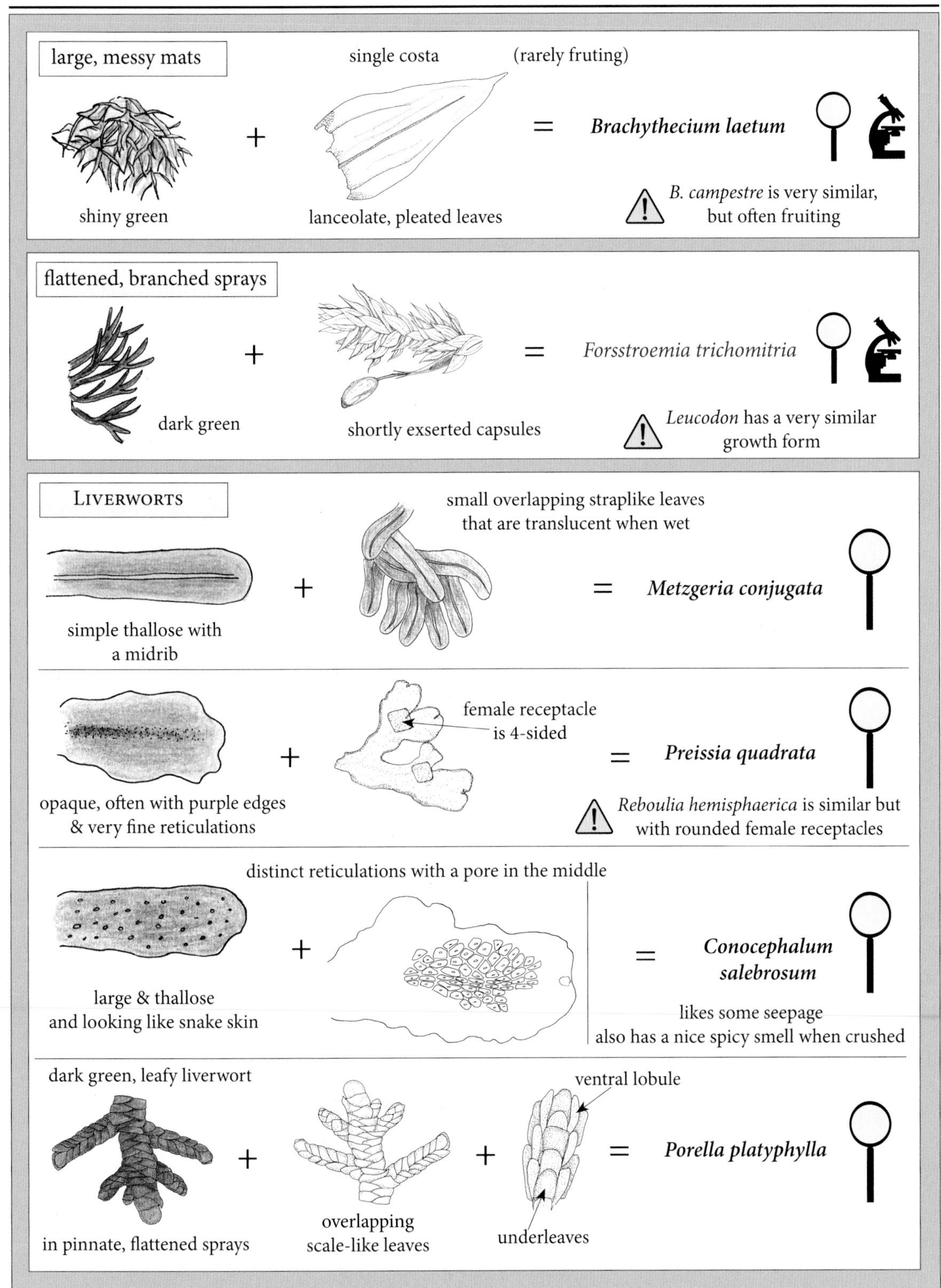
large, messy mats
single costa
(rarely fruting)
shiny green
lanceolate, pleated leaves
= Brachythecium laetum
B. campestre is very similar, but often fruiting
flattened, branched sprays
dark green
shortly exserted capsules
= Forsstroemia trichomitria
Leucodon has a very similar growth form
LIVERWORTS
small overlapping straplike leaves that are translucent when wet
simple thallose with a midrib
= Metzgeria conjugata
female receptacle is 4-sided
opaque, often with purple edges & very fine reticulations
= Preissia quadrata
Reboulia hemisphaerica is similar but with rounded female receptacles
distinct reticulations with a pore in the middle
large & thallose and looking like snake skin
= Conocephalum salebrosum
likes some seepage
also has a nice spicy smell when crushed
dark green, leafy liverwort
ventral lobule
in pinnate, flattened sprays
overlapping scale-like leaves
underleaves
= Porella platyphylla

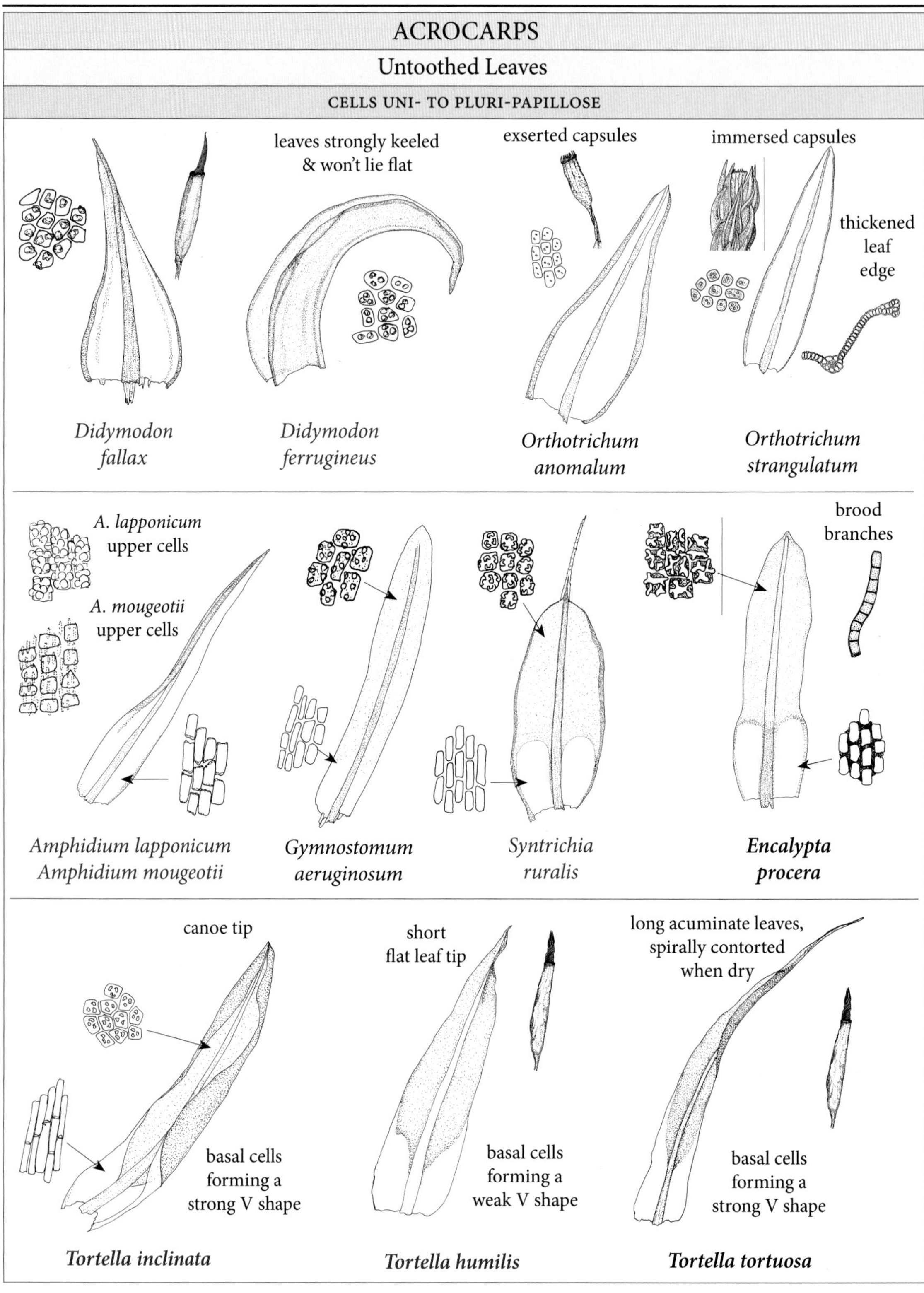
ACROCARPS
Untoothed Leaves
CELLS UNI- TO PLURI-PAPILLOSE
leaves strongly keeled & won't lie flat
exserted capsules
immersed capsules
thickened leaf edge
Didymodon fallax
Didymodon ferrugineus
Orthotrichum anomalum
Orthotrichum strangulatum
A. lapponicum upper cells
A. mougeotii upper cells
brood branches
Amphidium lapponicum
Amphidium mougeotii
Gymnostomum aeruginosum
Syntrichia ruralis
Encalypta procera
canoe tip
short flat leaf tip
long acuminate leaves, spirally contorted when dry
basal cells forming a strong V shape
basal cells forming a weak V shape
basal cells forming a strong V shape
Tortella inclinata
Tortella humilis
Tortella tortuosa

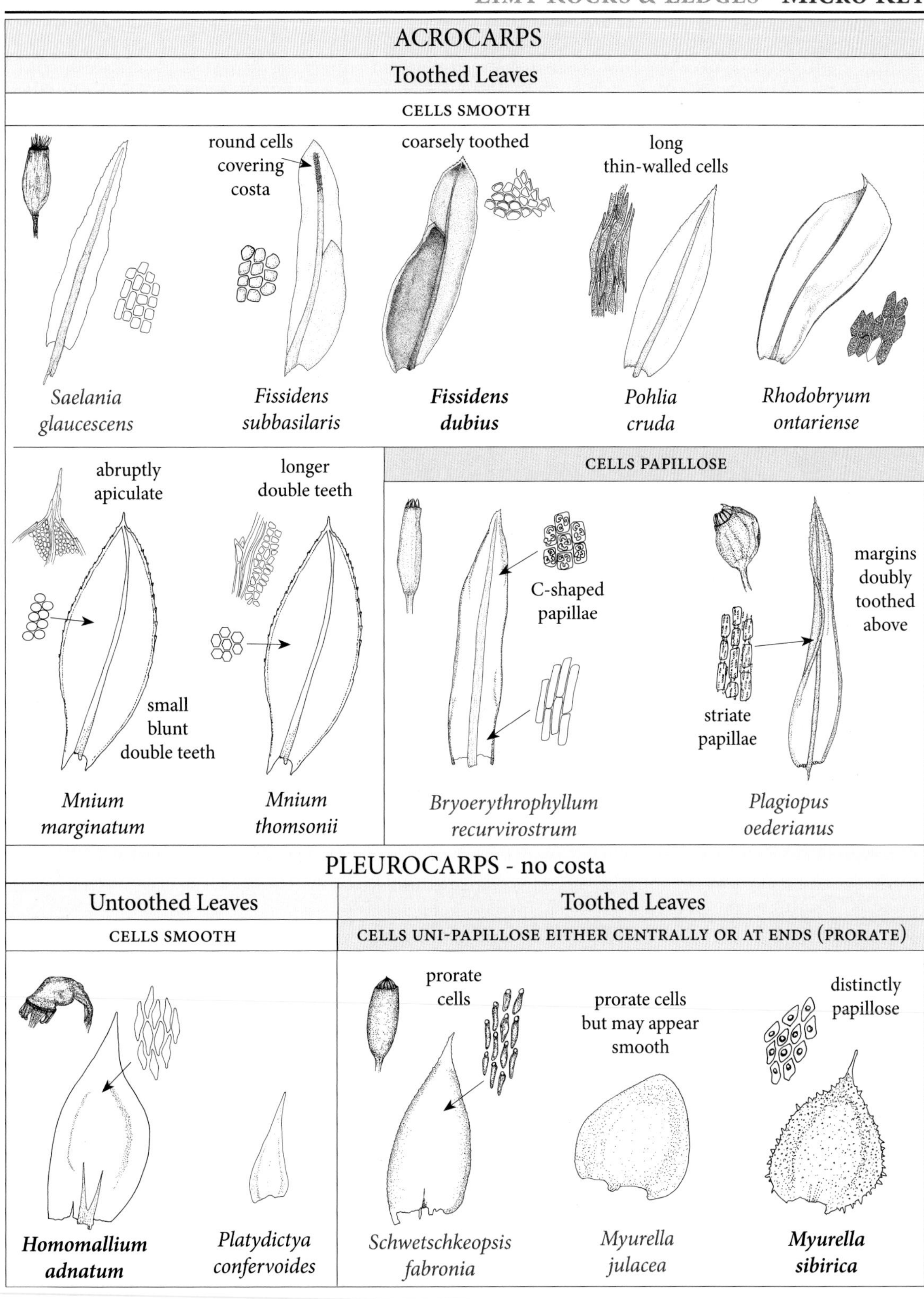
ACROCARPS
Toothed Leaves
CELLS SMOOTH
round cells covering costa
coarsely toothed
long thin-walled cells
Saelania glaucescens
Fissidens subbasilaris
Fissidens dubius
Pohlia cruda
Rhodobryum ontariense
abruptly apiculate
small blunt double teeth
longer double teeth
Mnium marginatum
Mnium thomsonii
CELLS PAPILLOSE
C-shaped papillae
margins doubly toothed above
striate papillae
Bryoerythrophyllum recurvirostrum
Plagiopus oederianus
PLEUROCARPS - no costa
Untoothed Leaves
Toothed Leaves
CELLS SMOOTH
CELLS UNI-PAPILLOSE EITHER CENTRALLY OR AT ENDS (PRORATE)
prorate cells
prorate cells but may appear smooth
distinctly papillose
Homomallium adnatum
Platydictya confervoides
Schwetschkeopsis fabronia
Myurella julacea
Myurella sibirica

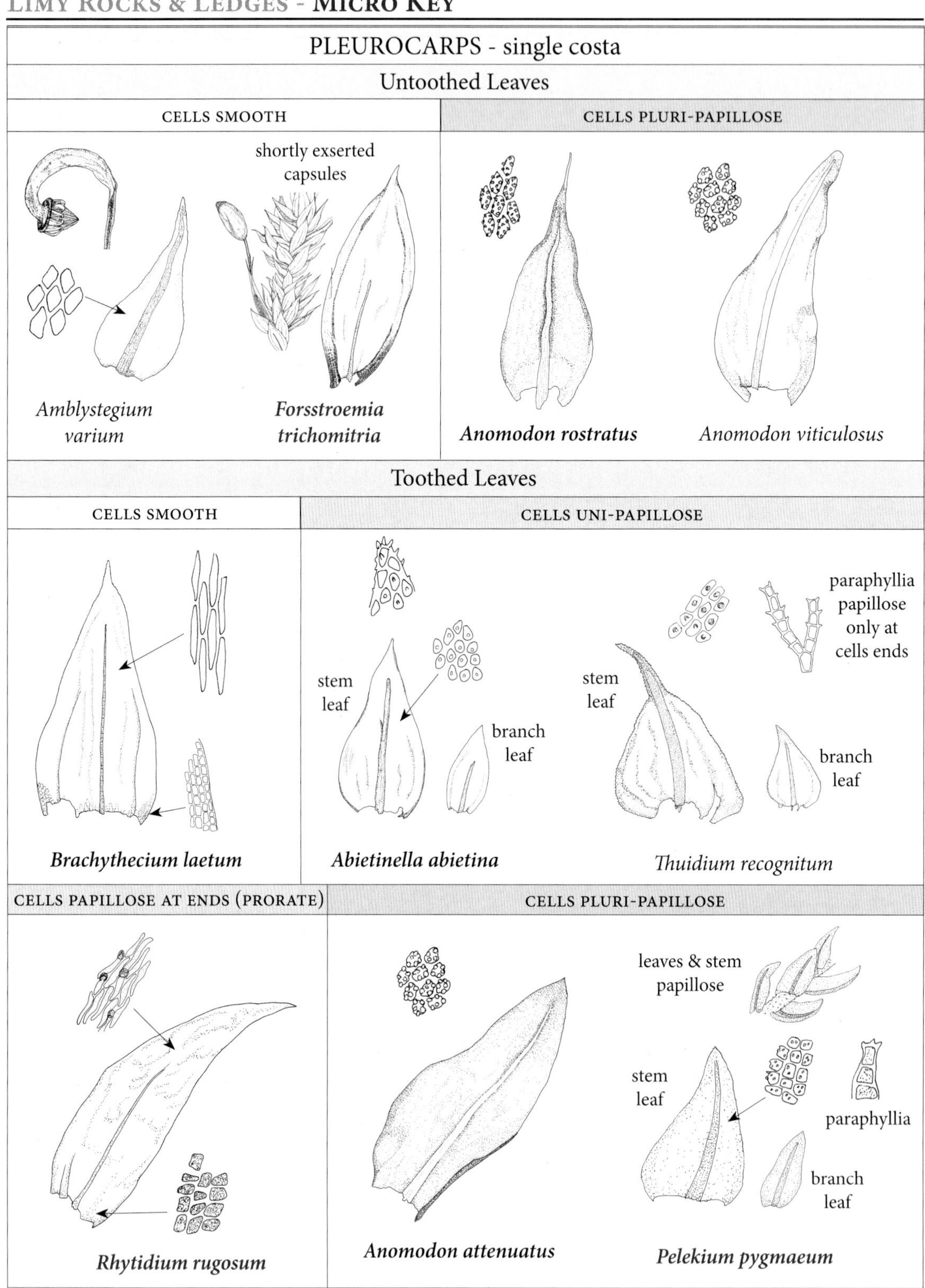
PLEUROCARPS - single costa
Untoothed Leaves
CELLS SMOOTH
CELLS PLURI-PAPILLOSE
shortly exserted
capsules
Amblystegium
varium
Forsstroemia
trichomitria
Anomodon rostratus
Anomodon viticulosus
Toothed Leaves
CELLS SMOOTH
CELLS UNI-PAPILLOSE
stem
leaf
branch
leaf
stem
leaf
paraphyllia
papillose
only at
cells ends
branch
leaf
Brachythecium laetum
Abietinella abietina
Thuidium recognitum
CELLS PAPILLOSE AT ENDS (PRORATE)
CELLS PLURI-PAPILLOSE
leaves & stem
papillose
stem
leaf
paraphyllia
branch
leaf
Rhytidium rugosum
Anomodon attenuatus
Pelekium pygmaeum

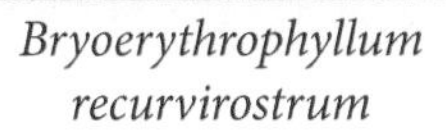

Bryoerythrophyllum recurvirostrum

Plagiopus oederianus

Amphidium lapponicum

Tortella inclinata

Tortella humilis

Tortella tortuosa

Encalypta procera

Syntrichia ruralis

Fissidens subbasilaris

Schwetschkeopsis fabronia

Myurella sibirica

Myurella julacea

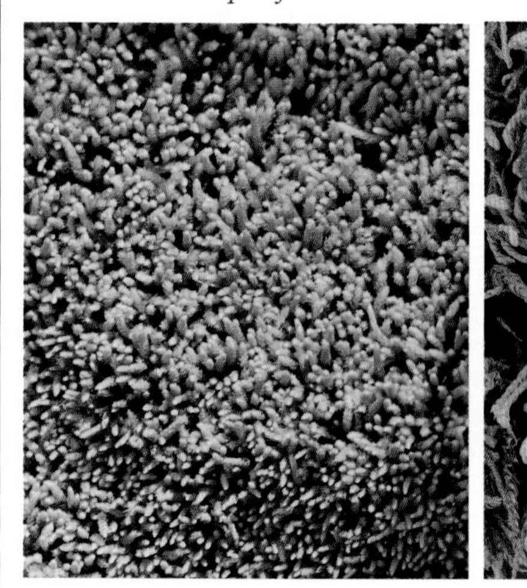

Anomodon rostratus

Anomodon attenuatus

Anomodon viticulosus

Abietinella abietina

Brachythecium laetum

Forsstroemia trichomitria

WET ROCKS & LEDGES

(SAW)

This section deals with those mosses that live on wet rocks, whether in a streambed, the splash zone of waterfalls or on ledges that have a more or less constant flow of water for at least part of the year.

Dry rock species can also be found in these areas on the rocks at the edges of the brook or the drier parts of the ledges, so be sure to check the previous two sections if you are unsure and nothing seems to fit. This section also includes those mosses that also need a little liminess to the rock, such as *Hyophila involuta*, *Dichodontium pellucidum* and *Amphidium*.

Some species prefer the lowest places in brooks, such as the *Fontinalis* group, and some liverworts, *Jubula pennsylvanica* and *Porella pinnata*, thereby remaining submerged as long as possible. Others are on the tops or sides of larger rocks and are submerged in the spring but might end up high and dry in the later months. The most common species on rocks or boulders mid-brook in this area seem to be *Hygrohypnum eugyrium*, *Torrentaria riparioides*, *Brachythecium plumosum* and *Hygroamblystegium tenax*.

Mosses that grow here have to be able to hang on tightly to the substrate to deal with the heavy water flows in the spring as well as scouring by sand and debris, not to mention ice in the winter. The mosses on ledges, although not having the water flow issue, do have to deal with ice stripping them off after winter. Many times you can see piles of moss at the base of a wet ledge in the spring.

(SAW)

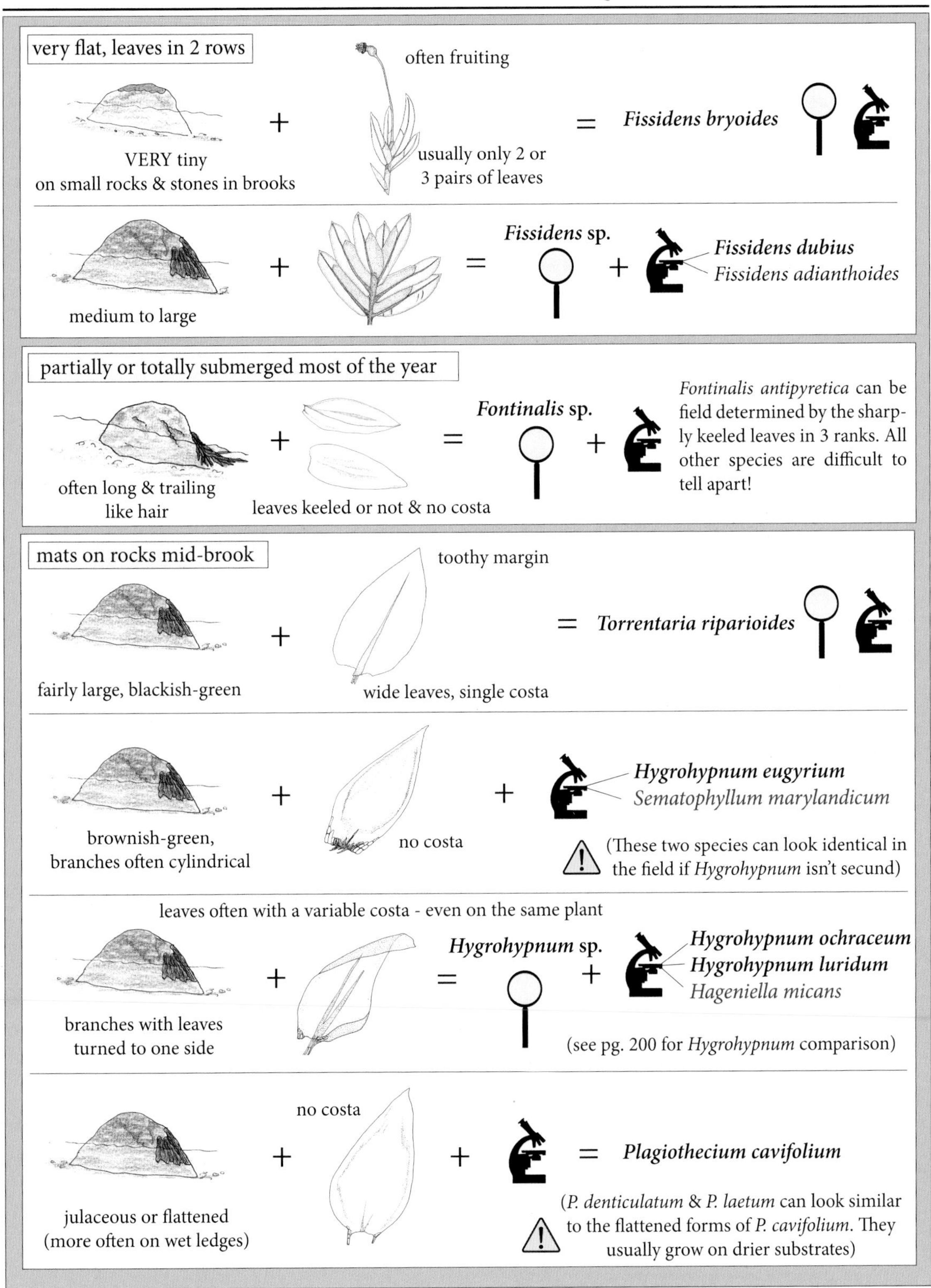

very flat, leaves in 2 rows
often fruiting
VERY tiny
on small rocks & stones in brooks
usually only 2 or
3 pairs of leaves
Fissidens bryoides
medium to large
Fissidens sp.
Fissidens dubius
Fissidens adianthoides
partially or totally submerged most of the year
often long & trailing
like hair
leaves keeled or not & no costa
Fontinalis sp.
Fontinalis antipyretica can be field determined by the sharply keeled leaves in 3 ranks. All other species are difficult to tell apart!
mats on rocks mid-brook
toothy margin
Torrentaria riparioides
fairly large, blackish-green
wide leaves, single costa
Hygrohypnum eugyrium
Sematophyllum marylandicum
brownish-green,
branches often cylindrical
no costa
(These two species can look identical in the field if Hygrohypnum isn't secund)
leaves often with a variable costa - even on the same plant
Hygrohypnum sp.
Hygrohypnum ochraceum
Hygrohypnum luridum
Hageniella micans
branches with leaves
turned to one side
(see pg. 200 for Hygrohypnum comparison)
no costa
Plagiothecium cavifolium
julaceous or flattened
(more often on wet ledges)
(P. denticulatum & P. laetum can look similar to the flattened forms of P. cavifolium. They usually grow on drier substrates)

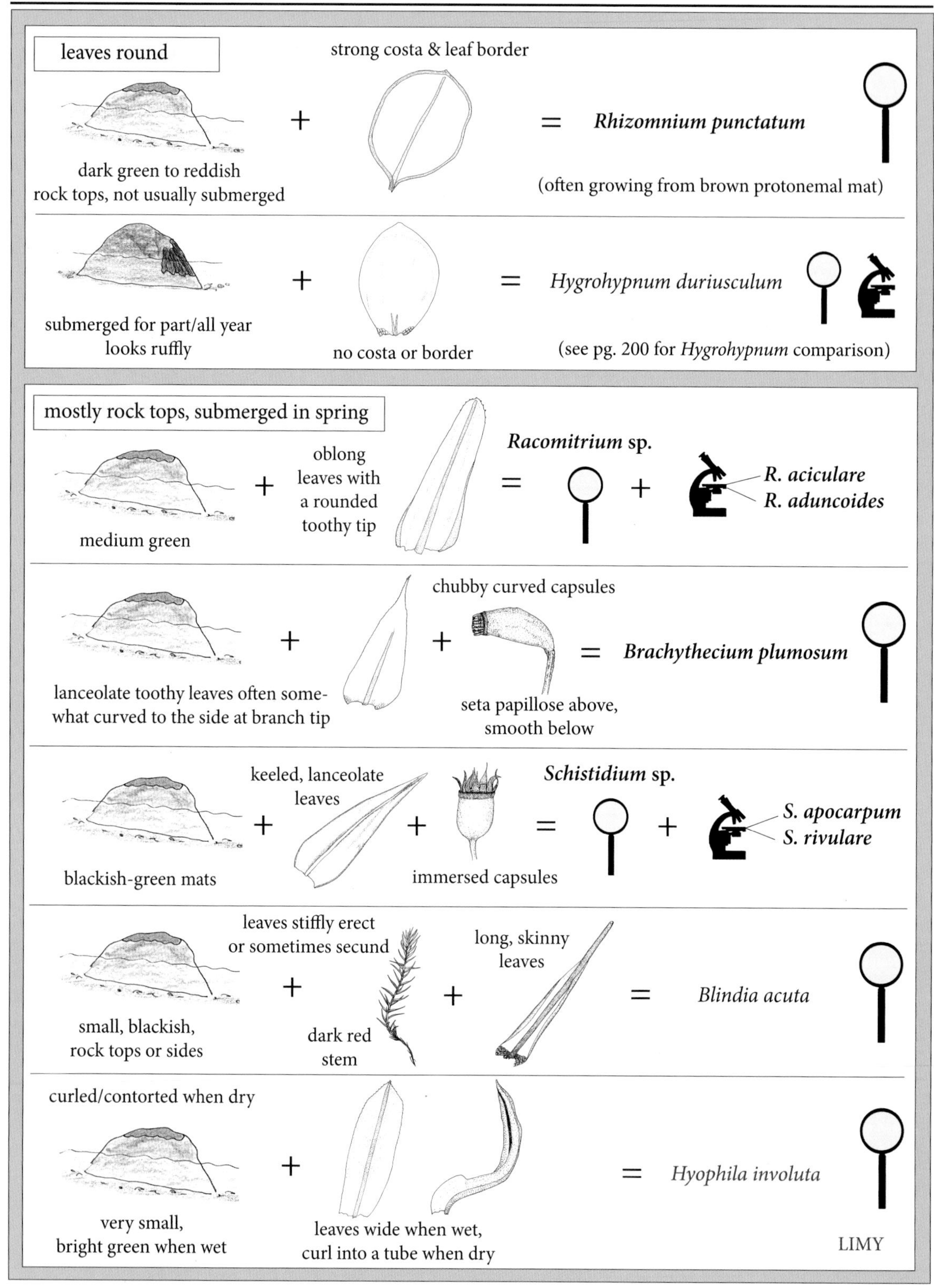
leaves round
strong costa & leaf border
dark green to reddish
rock tops, not usually submerged
+
=
Rhizomnium punctatum
(often growing from brown protonemal mat)
submerged for part/all year
looks ruffly
+
no costa or border
=
Hygrohypnum duriusculum
(see pg. 200 for *Hygrohypnum* comparison)
mostly rock tops, submerged in spring
medium green
+
oblong
leaves with
a rounded
toothy tip
Racomitrium sp.
=
+
R. aciculare
R. aduncoides
lanceolate toothy leaves often somewhat curved to the side at branch tip
+
+
chubby curved capsules
seta papillose above,
smooth below
=
Brachythecium plumosum
blackish-green mats
+
keeled, lanceolate
leaves
+
immersed capsules
Schistidium sp.
=
+
S. apocarpum
S. rivulare
small, blackish,
rock tops or sides
+
leaves stiffly erect
or sometimes secund
dark red
stem
+
long, skinny
leaves
=
Blindia acuta
curled/contorted when dry
very small,
bright green when wet
+
leaves wide when wet,
curl into a tube when dry
=
Hyophila involuta
LIMY

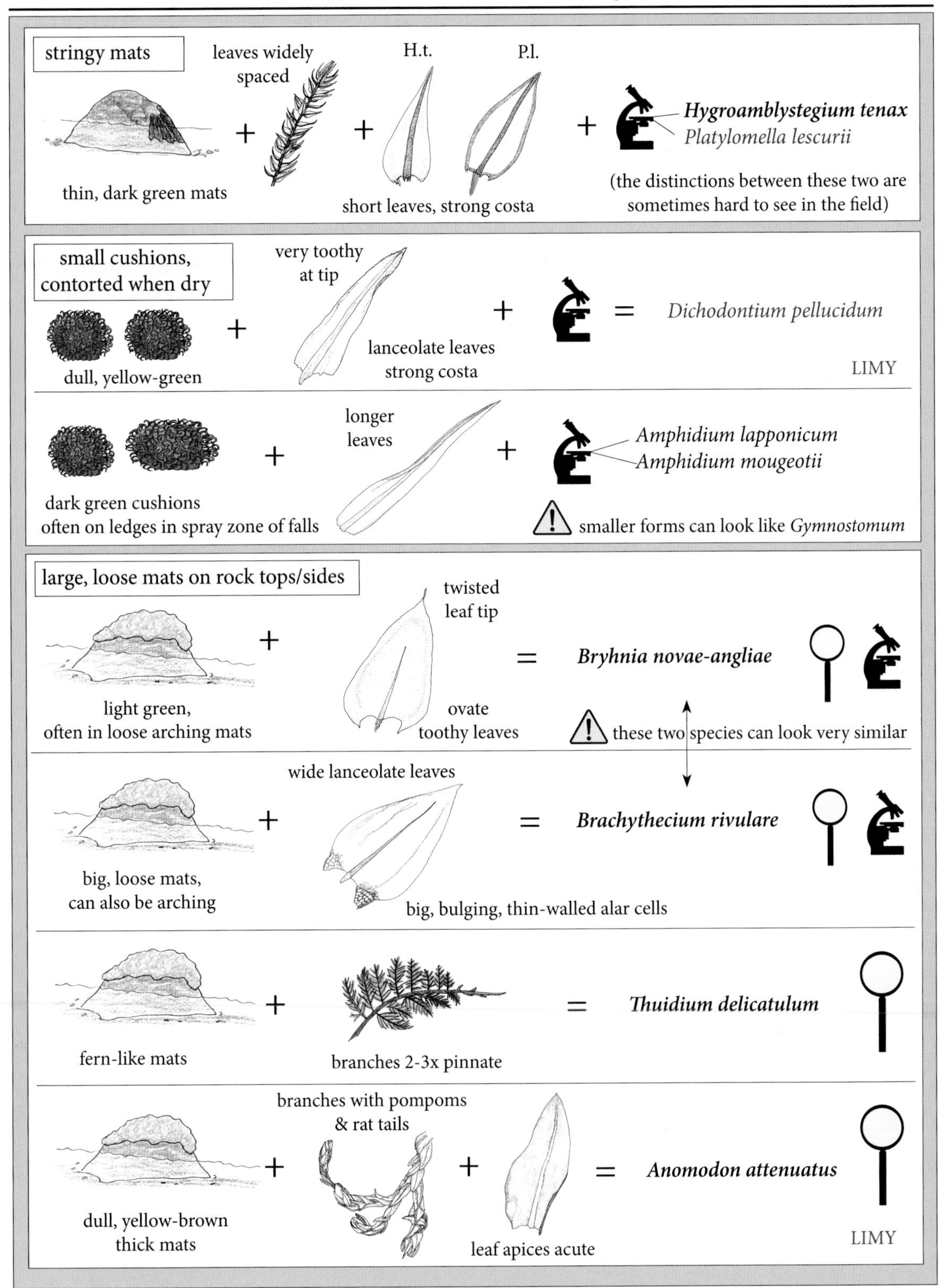
stringy mats
leaves widely spaced
H.t.
P.l.
+
+
+
Hygroamblystegium tenax
Platylomella lescurii
thin, dark green mats
short leaves, strong costa
(the distinctions between these two are sometimes hard to see in the field)
small cushions, contorted when dry
very toothy at tip
+
+
=
Dichodontium pellucidum
lanceolate leaves strong costa
dull, yellow-green
LIMY
longer leaves
+
+
Amphidium lapponicum
Amphidium mougeotii
dark green cushions
often on ledges in spray zone of falls
smaller forms can look like Gymnostomum
large, loose mats on rock tops/sides
twisted leaf tip
+
=
Bryhnia novae-angliae
light green,
often in loose arching mats
ovate toothy leaves
these two species can look very similar
wide lanceolate leaves
+
=
Brachythecium rivulare
big, loose mats,
can also be arching
big, bulging, thin-walled alar cells
+
=
Thuidium delicatulum
fern-like mats
branches 2-3x pinnate
branches with pompoms & rat tails
+
+
=
Anomodon attenuatus
dull, yellow-brown thick mats
leaf apices acute
LIMY

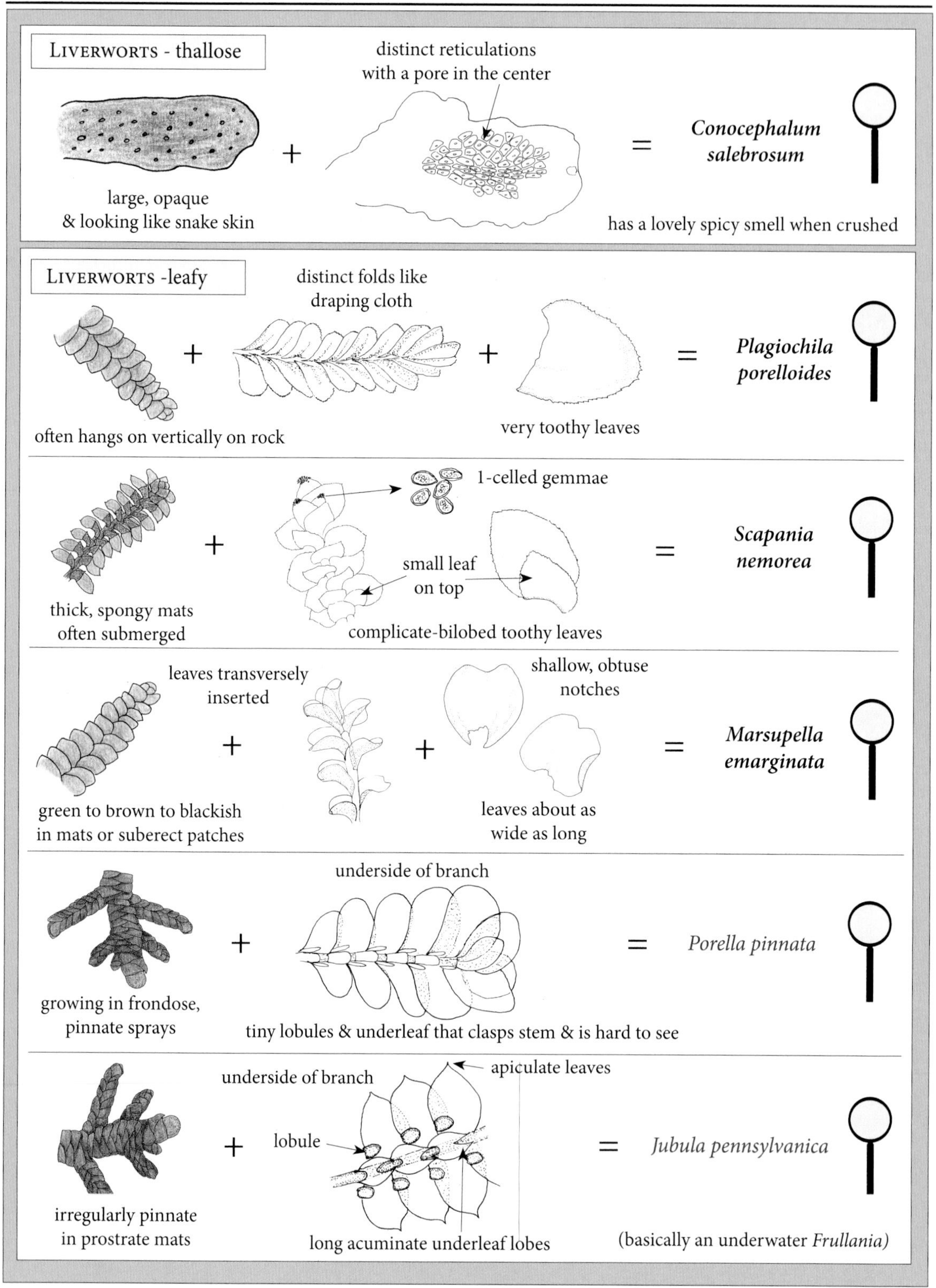
Liverworts - thallose
distinct reticulations
with a pore in the center
large, opaque
& looking like snake skin
+
=
Conocephalum
salebrosum
has a lovely spicy smell when crushed
Liverworts -leafy
distinct folds like
draping cloth
+
+
=
Plagiochila
porelloides
often hangs on vertically on rock
very toothy leaves
1-celled gemmae
+
small leaf
on top
=
Scapania
nemorea
thick, spongy mats
often submerged
complicate-bilobed toothy leaves
leaves transversely
inserted
shallow, obtuse
notches
+
+
=
Marsupella
emarginata
green to brown to blackish
in mats or suberect patches
leaves about as
wide as long
underside of branch
+
=
Porella pinnata
growing in frondose,
pinnate sprays
tiny lobules & underleaf that clasps stem & is hard to see
underside of branch
apiculate leaves
+
lobule
=
Jubula pennsylvanica
irregularly pinnate
in prostrate mats
long acuminate underleaf lobes
(basically an underwater Frullania)

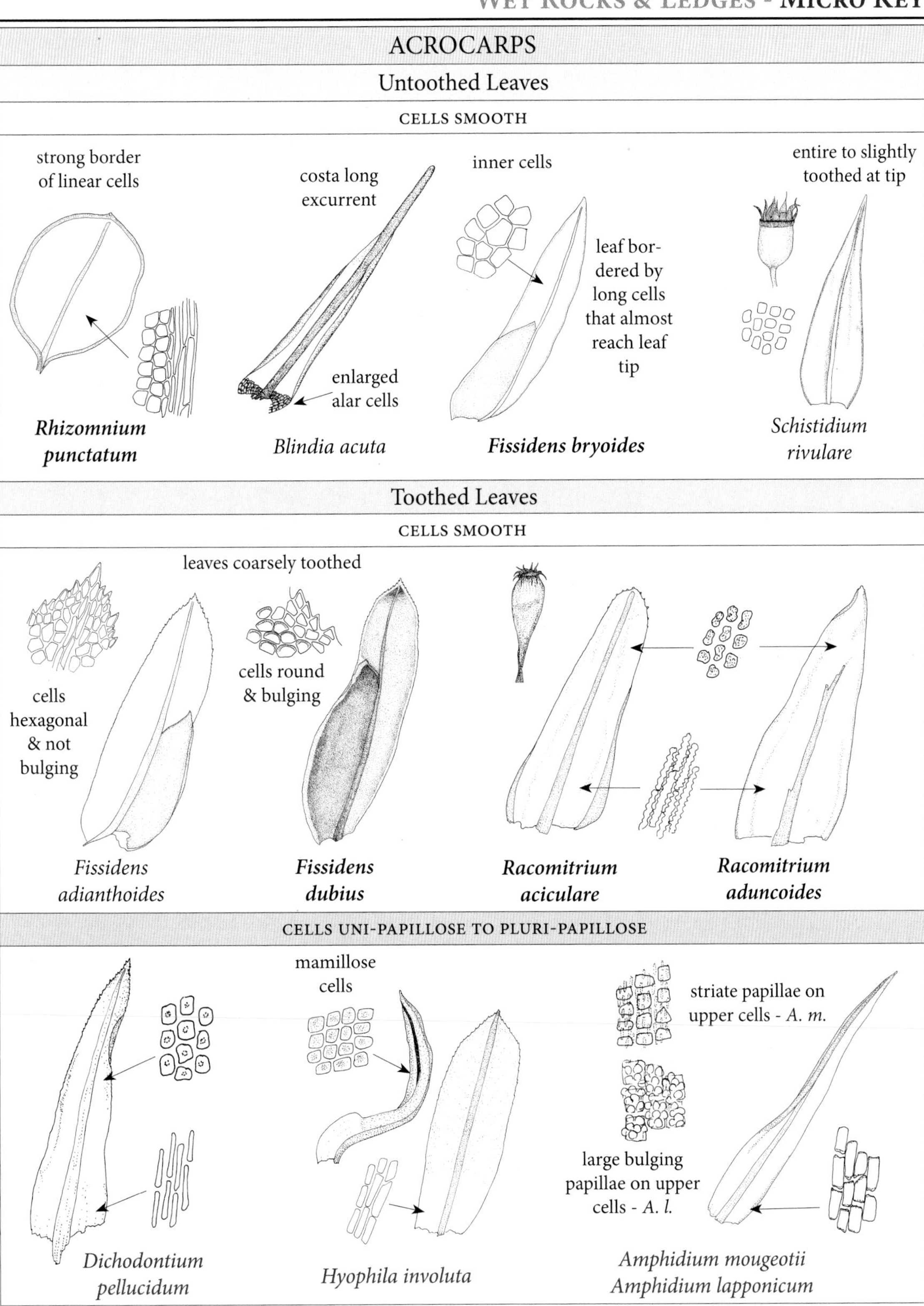
ACROCARPS
Untoothed Leaves
CELLS SMOOTH
strong border of linear cells
Rhizomnium punctatum
costa long excurrent
enlarged alar cells
Blindia acuta
inner cells
leaf bordered by long cells that almost reach leaf tip
Fissidens bryoides
entire to slightly toothed at tip
Schistidium rivulare
Toothed Leaves
CELLS SMOOTH
cells hexagonal & not bulging
Fissidens adianthoides
leaves coarsely toothed
cells round & bulging
Fissidens dubius
Racomitrium aciculare
Racomitrium aduncoides
CELLS UNI-PAPILLOSE TO PLURI-PAPILLOSE
Dichodontium pellucidum
mamillose cells
Hyophila involuta
striate papillae on upper cells - A. m.
large bulging papillae on upper cells - A. l.
Amphidium mougeotii
Amphidium lapponicum

PLEUROCARPS - double or no costa

Leaves Untoothed to Somewhat Toothed at Tip

CELLS SMOOTH

brood body
decurrencies
Plagiothecium cavifolium

strongly keeled leaves
Fontinalis antipyretica

Sematophyllum marylandicum

Hygrohypnum eugyrium

Hygrohypnum ochraceum

PLEUROCARPS - single costa

Leaves Untoothed to Somewhat Toothed at Tip

CELLS SMOOTH

Hygroamblystegium tenax

border
Platylomella lescurii

CELLS PLURI-PAPILLOSE

Anomodon attenuatus

Toothed Leaves

CELLS SMOOTH

cells shorter at tip
very toothy
Torrentaria riparioides

seta rough above, smooth below
Brachythecium plumosum

Brachythecium rivulare

CELLS UNI-PAPILLOSE

cells papillose at ends
Bryhnia novae-angliae

paraphyllia papillose at both cell middles and ends
papillae central
Thuidium delicatulum

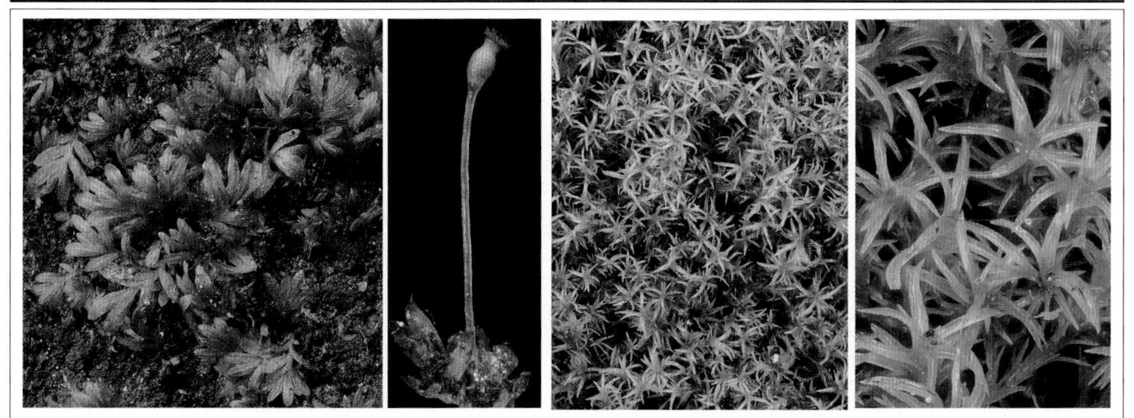

Fissidens bryoides *Amphidium mougeotii*

Fissidens adianthoides

Rhizomnium punctatum

Hyophila involuta

Racomitrium aduncoides

Racomitrium aciculare

Dichodontium pellucidum **_Bryhnia novae-angliae_**

Hygrohypnum eugyrium

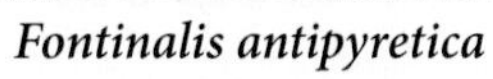

Fontinalis antipyretica

Platylomella lescurii

Hygroamblystegium tenax

WET SOIL

This section deals with mosses that live in places that stay pretty much consistently wet throughout the year. They can be found in both shaded areas such as wooded swamps or open areas such as pond shores, marshes or old beaver meadows. Some species are quite common and can be found in a wide variety of habitats. These include such species as *Aulacomnium palustre*, *Bryum pseudotriquetrum*, *Hypnum lindbergii*, *Philonotis fontana* and many of the Sphagnums.

Wooded swamps are usually dominated by a mixture of certain kinds of hardwoods and conifers, such as red maple and hemlock. There is little competition from herbaceous plants and shrubs, and these places often end up as root & pool swamps: drier mounds with pools in between and often dominated by *Sphagnum* species. The drier mounds will often have species that are commonly found on humic soil (see page 63). Likewise tree trunks, tree bases and logs have species found in those sections as well. The wetter parts of the mounds and the pools are where we're looking here. The pools can remain permanently flooded during the year or dry out towards the end of the season (in a dry year), and many of the pool species end up stranded on mud (*Leptodictyum riparium* & *Warnstorfia fluitans* are good examples). Another thing to note is that many of the wetland species (especially those growing in pools) are notoriously slow to get going in the spring. Spring *Bryhnia novae-angliae* looks much different from late summer *Bryhnia*! Not only branching can be different but coloration and tomentum development as well.

Favorite places to hunt for those species that like it wet but also like the open include not only pond shores but wet spots on powerline cuts, old beaver meadows and wet areas in gravel pits. Species that like these kinds of areas include *Atrichum crispum*, *Pohlia annotina* & *P. bulbifera* and *Sphagnum subsecundum*. These species don't tolerate competition from vascular plants well (such as grasses/sedges), whereas *Aulacomnium palustre*, *Bryum pseudotriquetrum* and *Hypnum lindbergii* can grow mixed in with these quite well.

Sphagnum dominated bogs have not been included here because there there are whole books devoted to *Sphagnum* identification. The *Sphagnum* species included are those that are common and/or fairly easy to identify in the field.

(SAW)

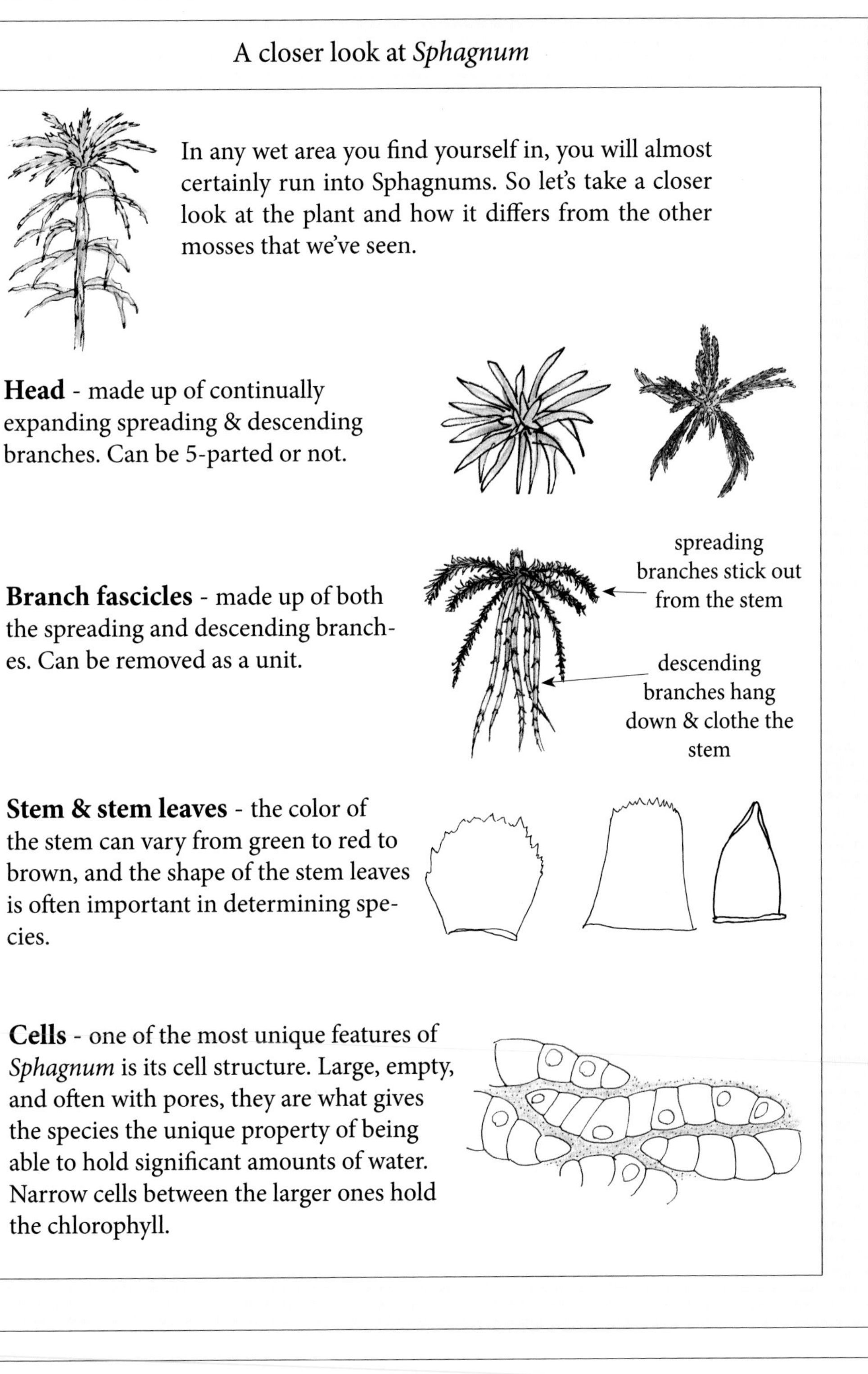

A closer look at *Sphagnum*

In any wet area you find yourself in, you will almost certainly run into Sphagnums. So let's take a closer look at the plant and how it differs from the other mosses that we've seen.

Head - made up of continually expanding spreading & descending branches. Can be 5-parted or not.

Branch fascicles - made up of both the spreading and descending branches. Can be removed as a unit.

Stem & stem leaves - the color of the stem can vary from green to red to brown, and the shape of the stem leaves is often important in determining species.

Cells - one of the most unique features of *Sphagnum* is its cell structure. Large, empty, and often with pores, they are what gives the species the unique property of being able to hold significant amounts of water. Narrow cells between the larger ones hold the chlorophyll.

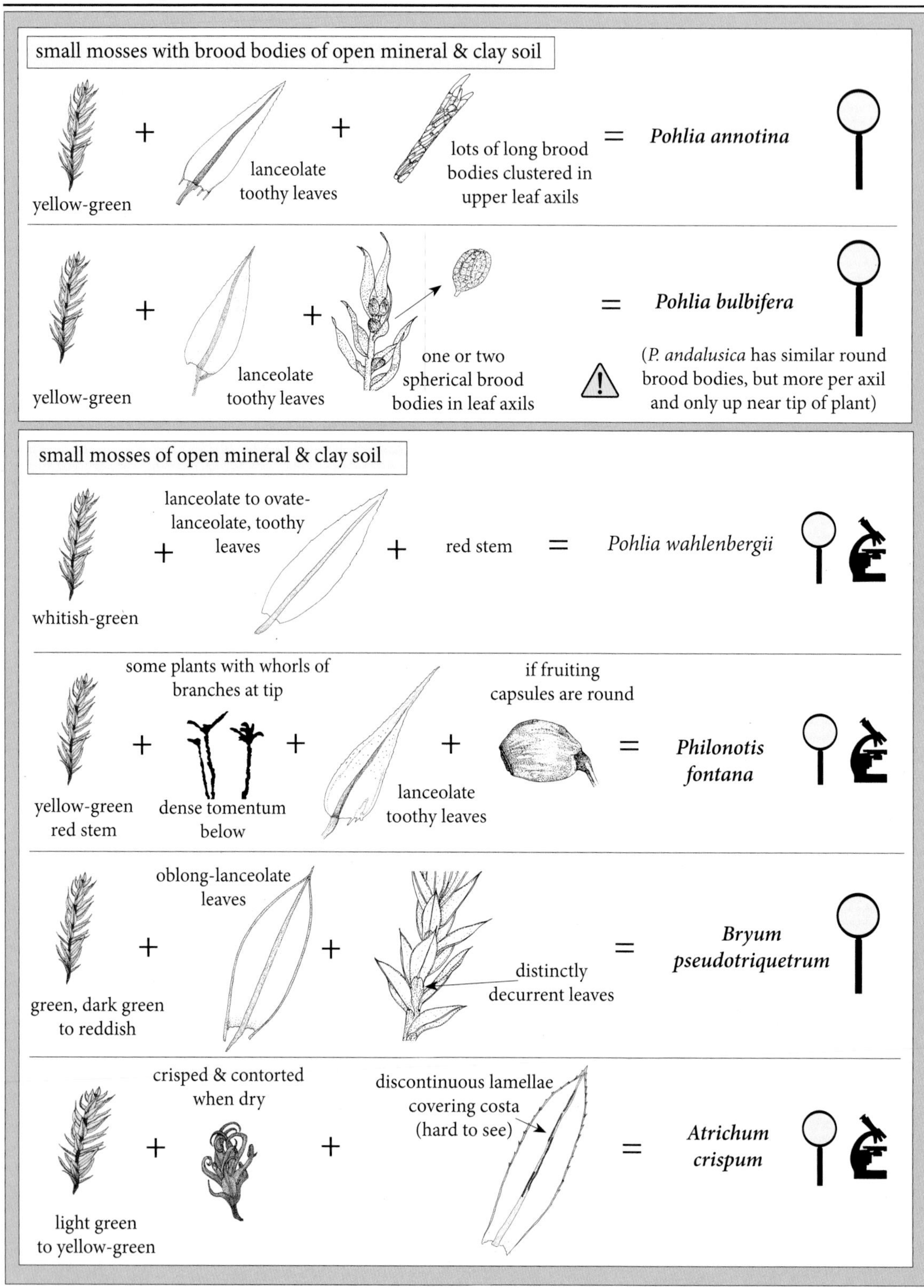
small mosses with brood bodies of open mineral & clay soil
yellow-green
+
lanceolate toothy leaves
+
lots of long brood bodies clustered in upper leaf axils
=
Pohlia annotina
yellow-green
+
lanceolate toothy leaves
+
one or two spherical brood bodies in leaf axils
=
Pohlia bulbifera
(P. andalusica has similar round brood bodies, but more per axil and only up near tip of plant)
small mosses of open mineral & clay soil
whitish-green
+
lanceolate to ovate-lanceolate, toothy leaves
+
red stem
=
Pohlia wahlenbergii
yellow-green red stem
+
some plants with whorls of branches at tip
dense tomentum below
+
lanceolate toothy leaves
+
if fruiting capsules are round
=
Philonotis fontana
green, dark green to reddish
+
oblong-lanceolate leaves
+
distinctly decurrent leaves
=
Bryum pseudotriquetrum
light green to yellow-green
+
crisped & contorted when dry
+
discontinuous lamellae covering costa (hard to see)
=
Atrichum crispum

long leaves, crisped & contorted when dry

medium to large plants + costa covered by ridges (lamellae); lamellae much less than 1/3 of the leaf blade; leaves wavy = ***Atrichum undulatum s.l.***

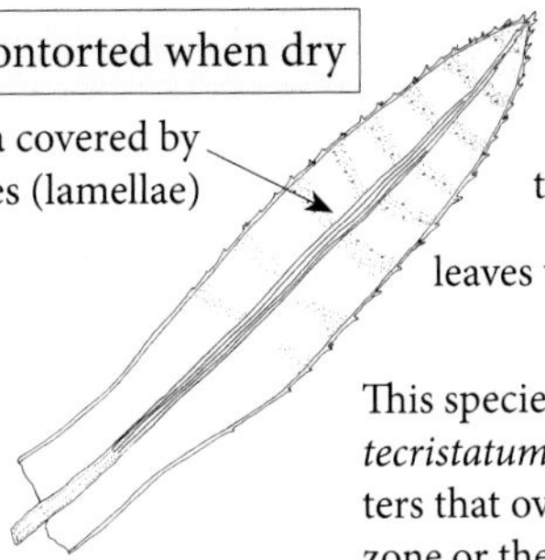

(small forms of this can look very much like *A. angustatum*)

This species has been split into 3 species by some authors: *A. altecristatum*, *A. crispulum* & *A. undulatum*. They all have characters that overlap and invariably you come up within the overlap zone or they have characters that fit one species & other characters that fit another. It's been left here in the broadest sense. See page 184 for *Atrichum* comparison.

yellowish with lots of brown tomentum on stem + lanceolate leaves often widely spaced + often with seta-like extension with leaf-shaped gemmae on top = ***Aulacomnium palustre***

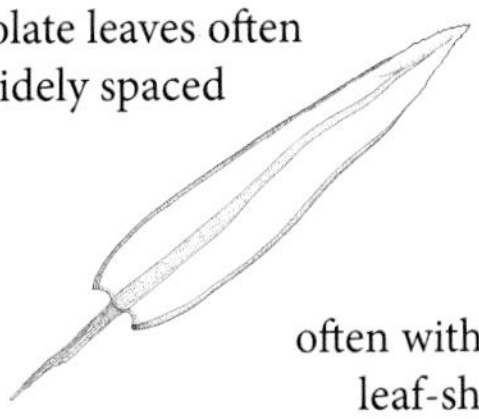

toothy leaves bordered by long cells

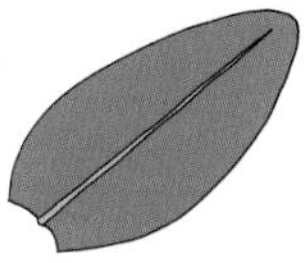

large elliptical leaves 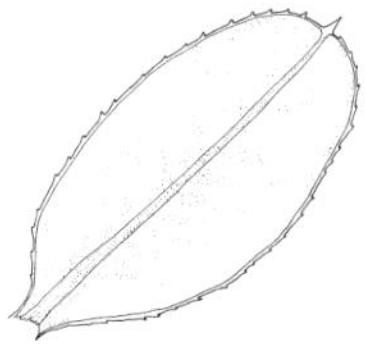 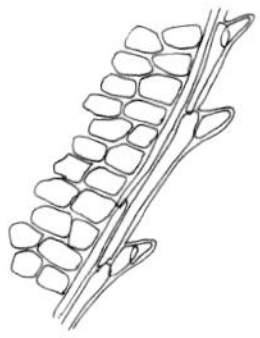 + mostly single-cell teeth all the way to base = ***Plagiomnium ellipticum***

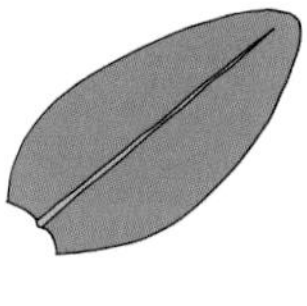

large elliptical to oblong-ovate leaves 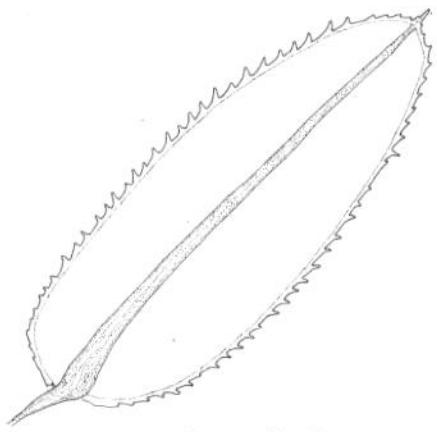 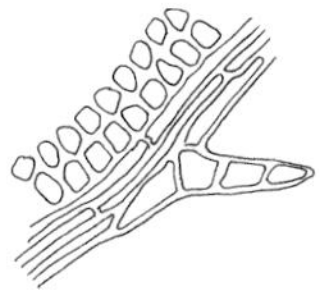 + multicellular teeth all the way to base = ***Plagiomnium ciliare***

bright green new leaves in spring contrasts with dark dirty green old leaves

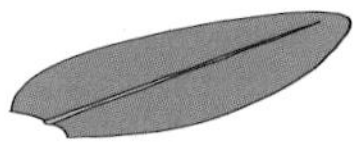

narrower double-toothed leaves 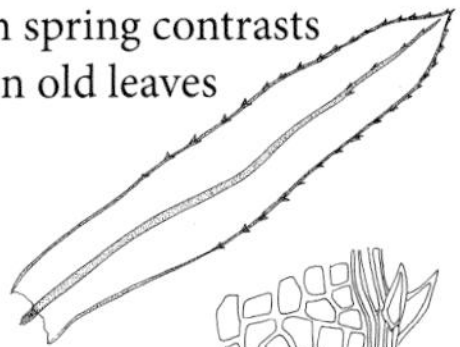 + sharp double teeth 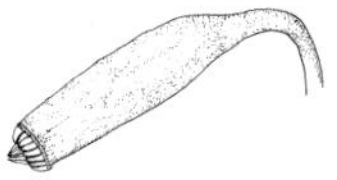 + droopy capsules (when found) = ***Mnium hornum***

(especially fond of stream banks)

Wet Soil - Quick & Dirty Field Key

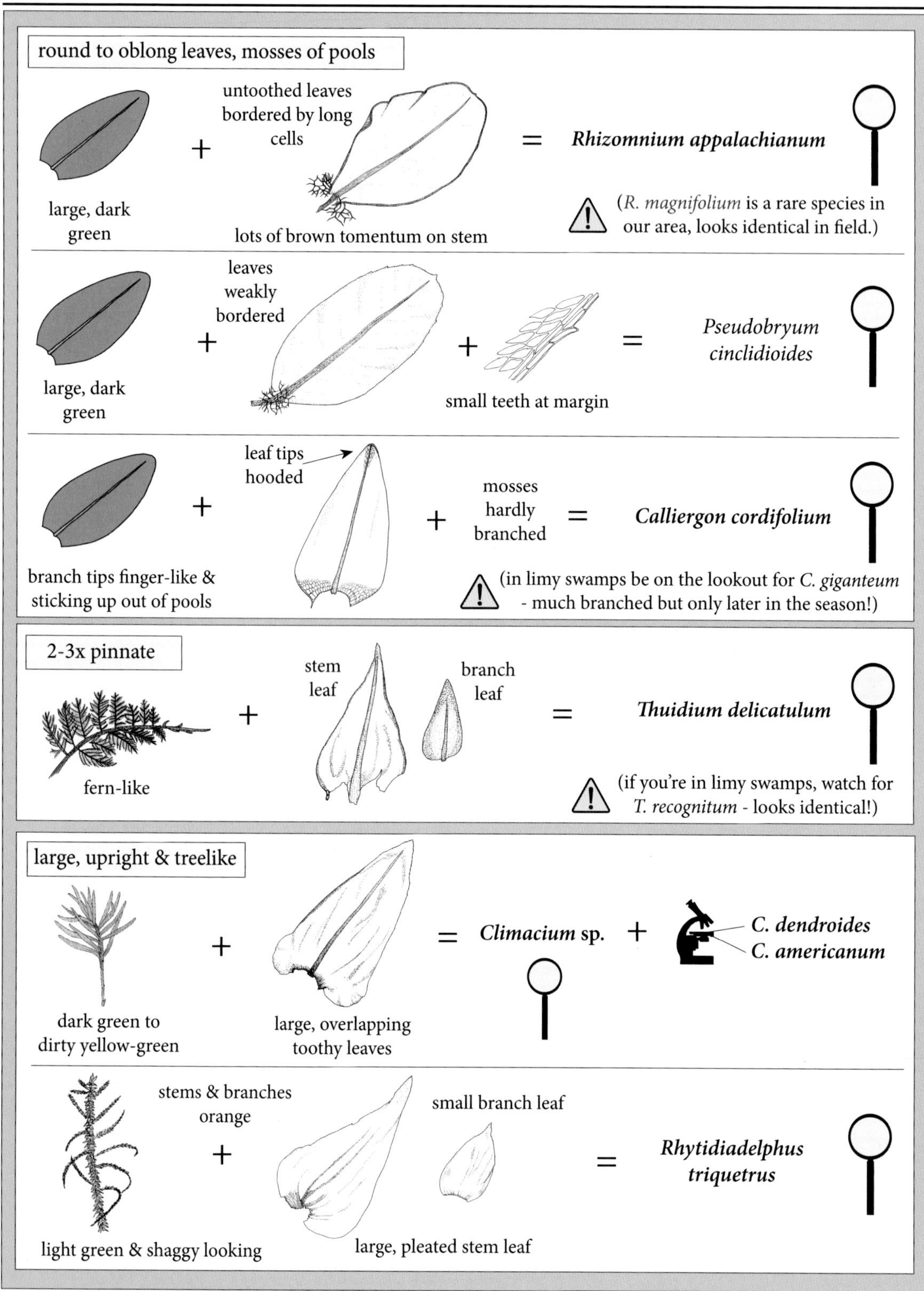

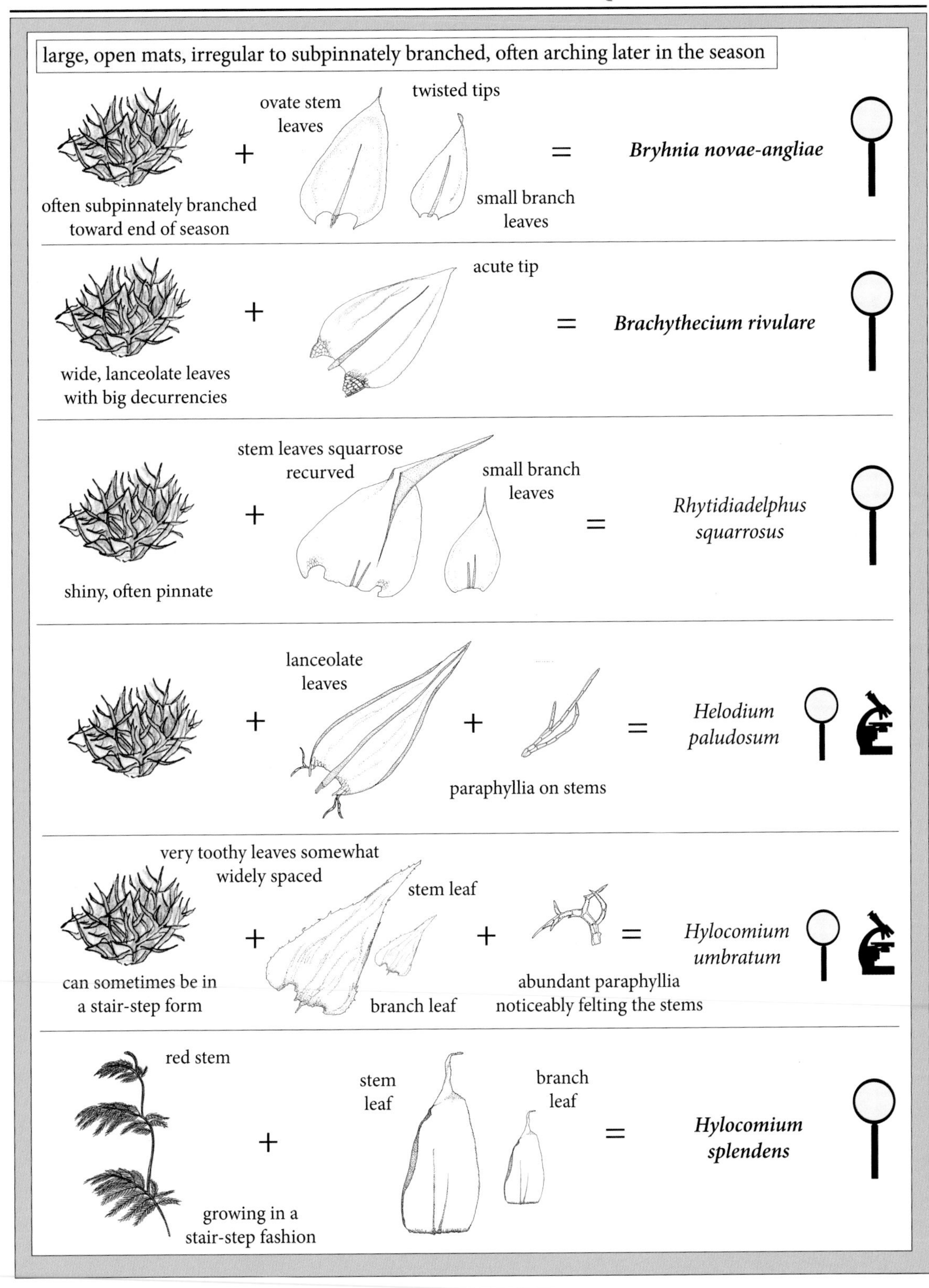
large, open mats, irregular to subpinnately branched, often arching later in the season
ovate stem leaves
twisted tips
small branch leaves
often subpinnately branched toward end of season
Bryhnia novae-angliae
acute tip
wide, lanceolate leaves with big decurrencies
Brachythecium rivulare
stem leaves squarrose recurved
small branch leaves
shiny, often pinnate
Rhytidiadelphus squarrosus
lanceolate leaves
paraphyllia on stems
Helodium paludosum
very toothy leaves somewhat widely spaced
stem leaf
branch leaf
can sometimes be in a stair-step form
abundant paraphyllia noticeably felting the stems
Hylocomium umbratum
red stem
growing in a stair-step fashion
stem leaf
branch leaf
Hylocomium splendens

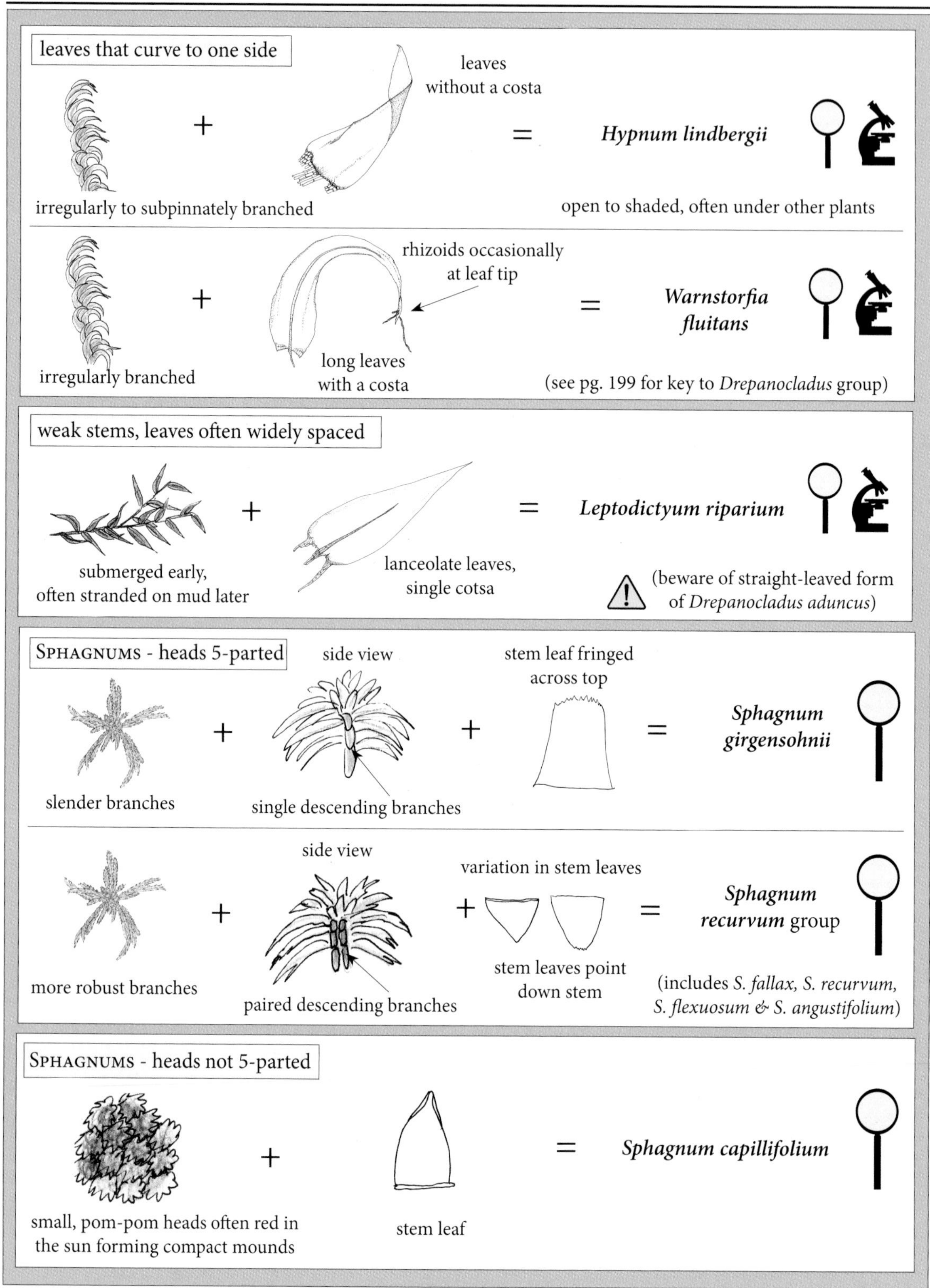
leaves that curve to one side
irregularly to subpinnately branched
leaves without a costa
Hypnum lindbergii
open to shaded, often under other plants
irregularly branched
rhizoids occasionally at leaf tip
long leaves with a costa
Warnstorfia fluitans
(see pg. 199 for key to Drepanocladus group)
weak stems, leaves often widely spaced
submerged early, often stranded on mud later
lanceolate leaves, single cotsa
Leptodictyum riparium
(beware of straight-leaved form of Drepanocladus aduncus)
Sphagnums - heads 5-parted
slender branches
side view
single descending branches
stem leaf fringed across top
Sphagnum girgensohnii
more robust branches
side view
paired descending branches
variation in stem leaves
stem leaves point down stem
Sphagnum recurvum group
(includes S. fallax, S. recurvum, S. flexuosum & S. angustifolium)
Sphagnums - heads not 5-parted
small, pom-pom heads often red in the sun forming compact mounds
stem leaf
Sphagnum capillifolium

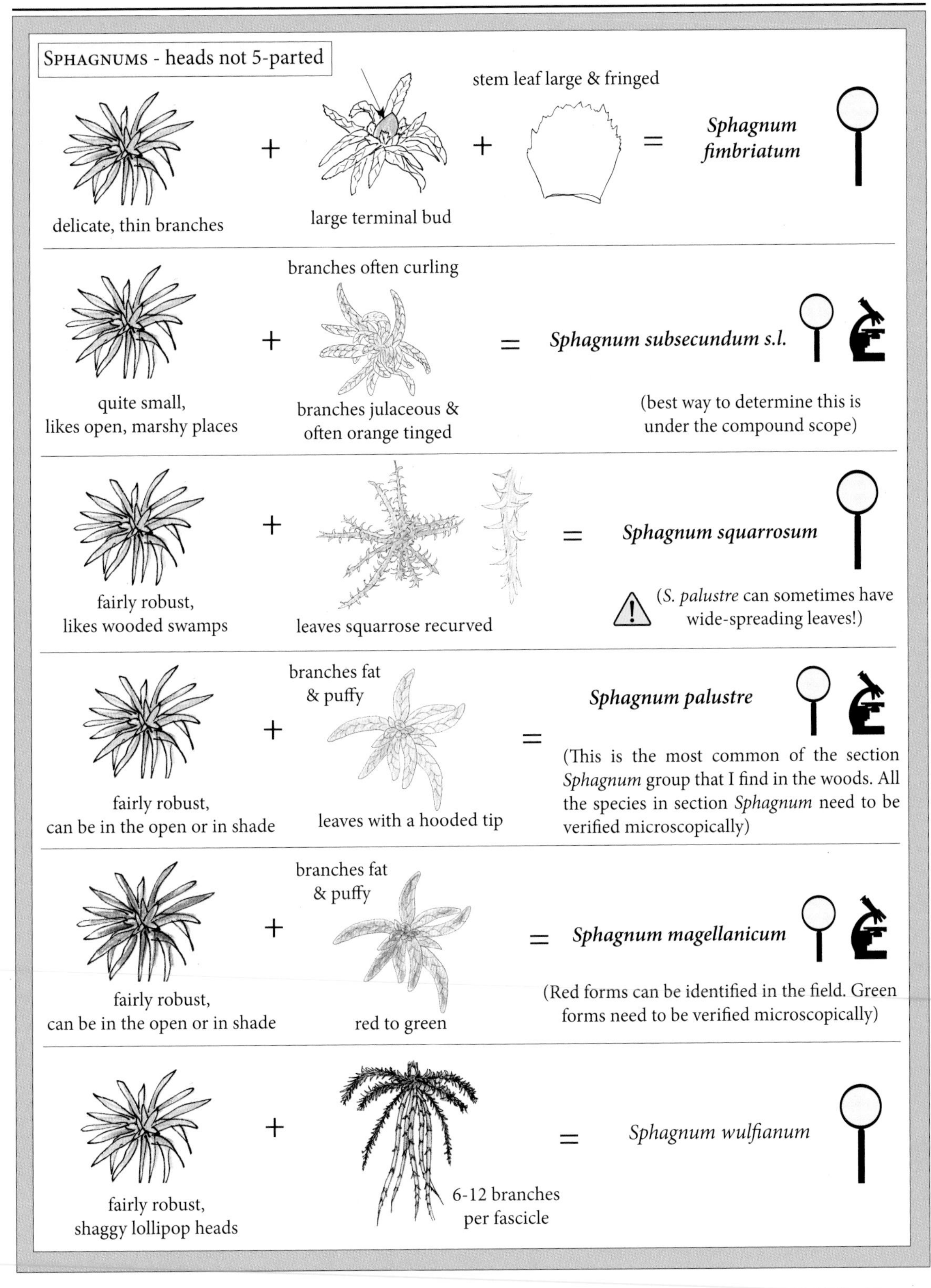
Sphagnums - heads not 5-parted
stem leaf large & fringed
delicate, thin branches
+
large terminal bud
+
=
Sphagnum fimbriatum
branches often curling
quite small,
likes open, marshy places
+
branches julaceous &
often orange tinged
=
Sphagnum subsecundum s.l.
(best way to determine this is
under the compound scope)
fairly robust,
likes wooded swamps
+
leaves squarrose recurved
=
Sphagnum squarrosum
(S. palustre can sometimes have
wide-spreading leaves!)
branches fat
& puffy
fairly robust,
can be in the open or in shade
+
leaves with a hooded tip
=
Sphagnum palustre
(This is the most common of the section Sphagnum group that I find in the woods. All the species in section Sphagnum need to be verified microscopically)
branches fat
& puffy
fairly robust,
can be in the open or in shade
+
red to green
=
Sphagnum magellanicum
(Red forms can be identified in the field. Green forms need to be verified microscopically)
fairly robust,
shaggy lollipop heads
+
6-12 branches
per fascicle
=
Sphagnum wulfianum

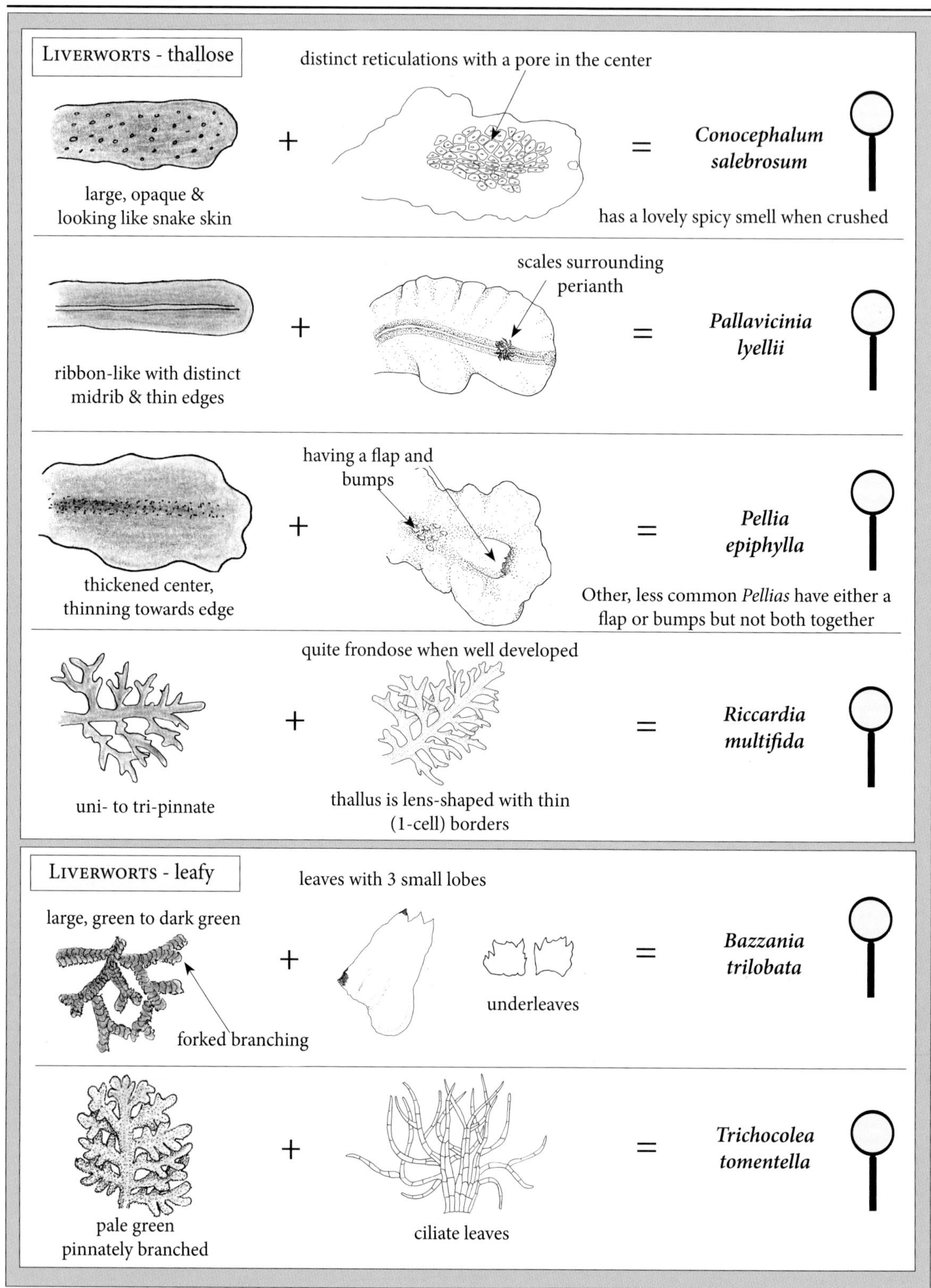
Liverworts - thallose
distinct reticulations with a pore in the center
+
=
Conocephalum salebrosum
large, opaque & looking like snake skin
has a lovely spicy smell when crushed
scales surrounding perianth
+
=
Pallavicinia lyellii
ribbon-like with distinct midrib & thin edges
having a flap and bumps
+
=
Pellia epiphylla
thickened center, thinning towards edge
Other, less common Pellias have either a flap or bumps but not both together
quite frondose when well developed
+
=
Riccardia multifida
uni- to tri-pinnate
thallus is lens-shaped with thin (1-cell) borders
Liverworts - leafy
leaves with 3 small lobes
large, green to dark green
+
=
Bazzania trilobata
underleaves
forked branching
+
=
Trichocolea tomentella
pale green pinnately branched
ciliate leaves

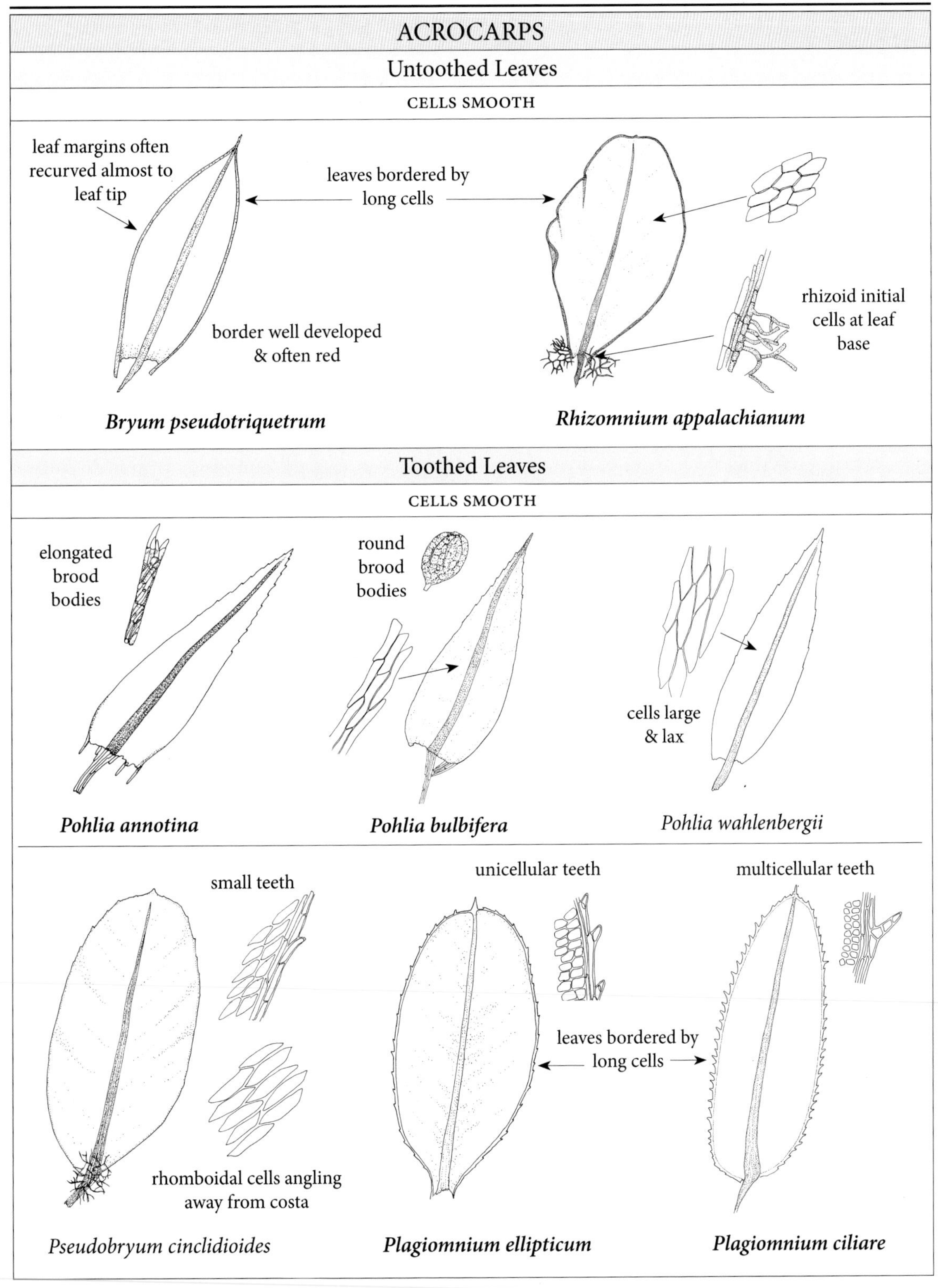
ACROCARPS
Untoothed Leaves
CELLS SMOOTH
leaf margins often recurved almost to leaf tip
leaves bordered by long cells
border well developed & often red
rhizoid initial cells at leaf base
Bryum pseudotriquetrum
Rhizomnium appalachianum
Toothed Leaves
CELLS SMOOTH
elongated brood bodies
round brood bodies
cells large & lax
Pohlia annotina
Pohlia bulbifera
Pohlia wahlenbergii
small teeth
unicellular teeth
multicellular teeth
leaves bordered by long cells
rhomboidal cells angling away from costa
Pseudobryum cinclidioides
Plagiomnium ellipticum
Plagiomnium ciliare

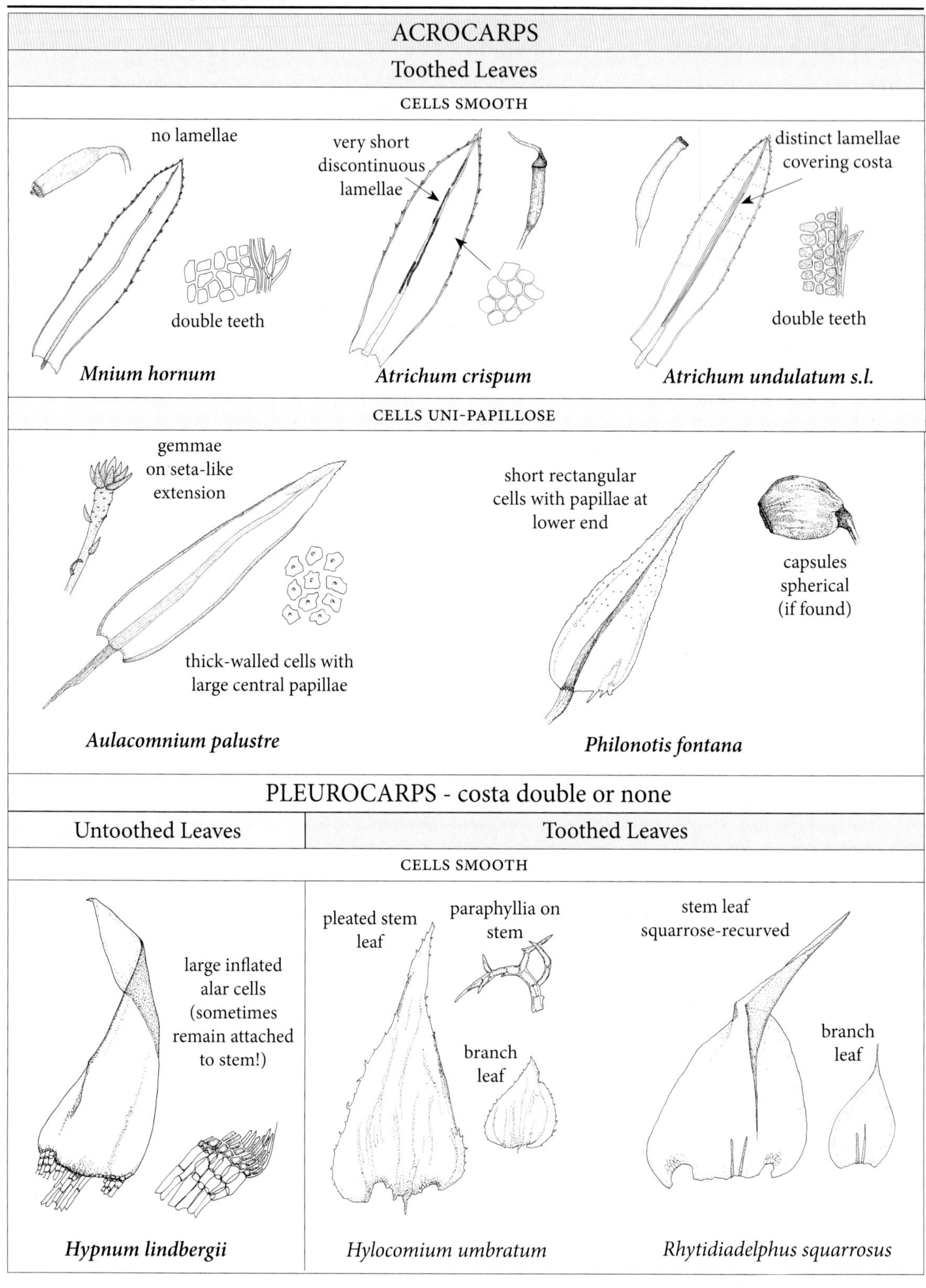
ACROCARPS
Toothed Leaves
CELLS SMOOTH
no lamellae
double teeth
Mnium hornum
very short discontinuous lamellae
Atrichum crispum
distinct lamellae covering costa
double teeth
Atrichum undulatum s.l.
CELLS UNI-PAPILLOSE
gemmae on seta-like extension
thick-walled cells with large central papillae
Aulacomnium palustre
short rectangular cells with papillae at lower end
capsules spherical (if found)
Philonotis fontana
PLEUROCARPS - costa double or none
Untoothed Leaves
Toothed Leaves
CELLS SMOOTH
large inflated alar cells (sometimes remain attached to stem!)
Hypnum lindbergii
pleated stem leaf
paraphyllia on stem
branch leaf
Hylocomium umbratum
stem leaf squarrose-recurved
branch leaf
Rhytidiadelphus squarrosus

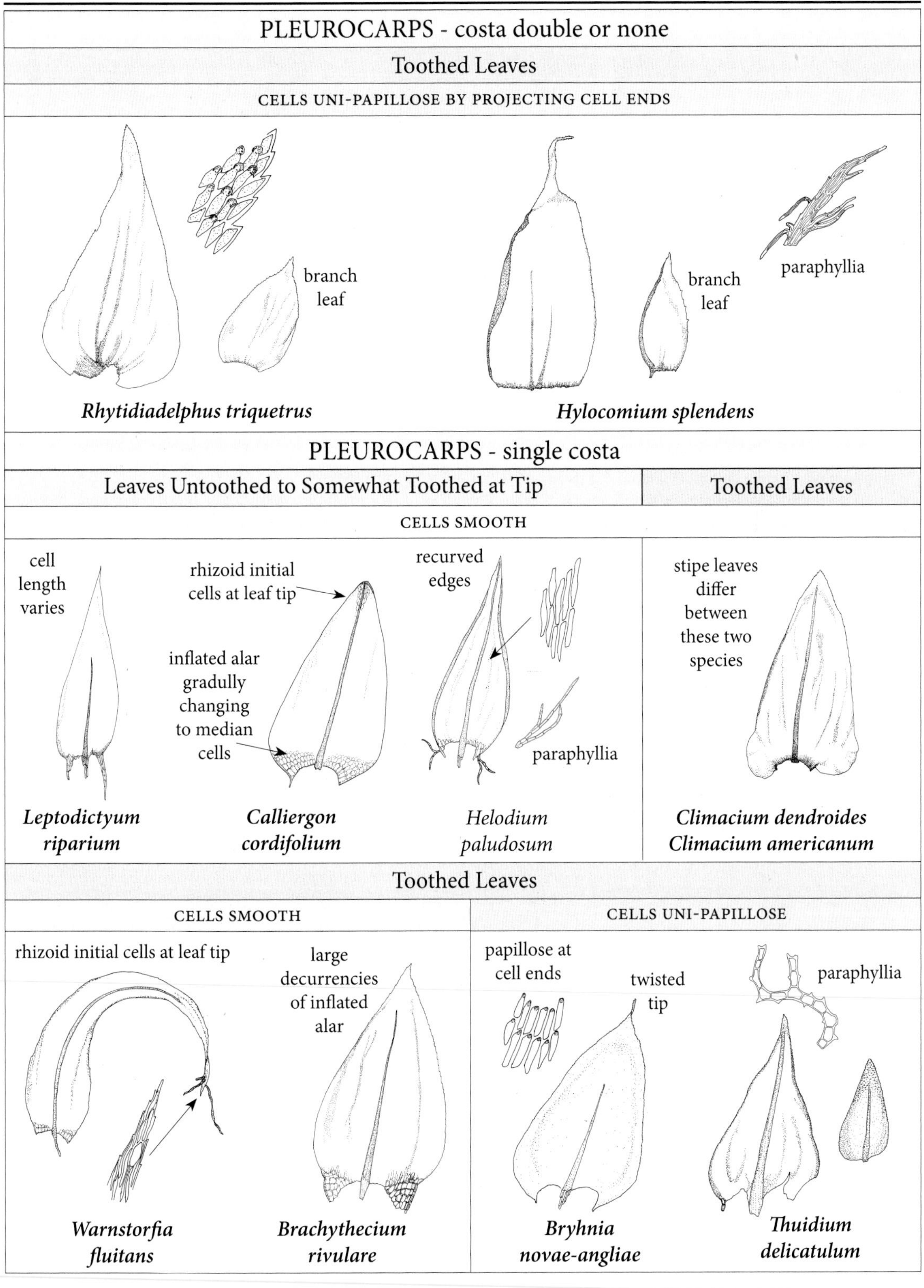
PLEUROCARPS - costa double or none
Toothed Leaves
CELLS UNI-PAPILLOSE BY PROJECTING CELL ENDS
branch leaf
Rhytidiadelphus triquetrus
branch leaf
paraphyllia
Hylocomium splendens
PLEUROCARPS - single costa
Leaves Untoothed to Somewhat Toothed at Tip
Toothed Leaves
CELLS SMOOTH
cell length varies
Leptodictyum riparium
rhizoid initial cells at leaf tip
inflated alar gradully changing to median cells
Calliergon cordifolium
recurved edges
paraphyllia
Helodium paludosum
stipe leaves differ between these two species
Climacium dendroides
Climacium americanum
Toothed Leaves
CELLS SMOOTH
CELLS UNI-PAPILLOSE
rhizoid initial cells at leaf tip
Warnstorfia fluitans
large decurrencies of inflated alar
Brachythecium rivulare
papillose at cell ends
twisted tip
Bryhnia novae-angliae
paraphyllia
Thuidium delicatulum

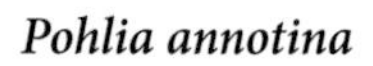

Pohlia annotina

Pohlia bulbifera

Aulacomnium palustre

Rhizomnium appalachianum

Pseudobryum cinclidioides

Philonotis fontana

Hypnum lindbergii

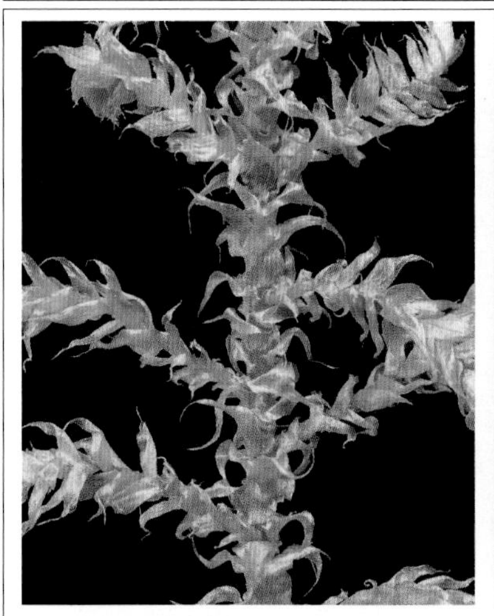

Rhytidiadelphus squarrosus

Rhytidiadelphus triquetrus

Climacium dendroides

Calliergon cordifolium

Bryum pseudotriquetrum

Sphagnum squarrosum

Sphagnum wulfianum

Sphagnum capillifolium

Sphagnum palustre

Sphagnum recurvum group

Conocephalum salebrosum

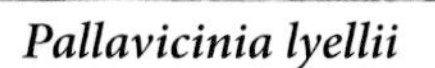

Pallavicinia lyellii SAW

Pellia ephiphylla SAW

Trichocolea tomentella SAW

RICH FENS & LIMY SEEPS

Rich fens & limy seeps are a type of open wetland where nutrients are higher than in bogs or wooded swamps; they are also more influenced by groundwater. Fens, especially, can be dominated by shrubs, grasses & sedges as well as mosses.

The mosses found here are often referred to as 'brown mosses', and they include such species as *Calliergonella cuspidata, Cratoneuron filicinum, Drepanocladus aduncus, Limprichtia revolvens, Scorpidium scorpioides* and *Tomentypnum nitens* as well as some of the Sphagnums.

This sections contains bryophytes that are fairly restricted to this type of habitat. But they are not the only mosses you will find. Several of the species listed in the general wetlands section (pages 145-160) will be found here as well, such as *Aulacomnium palustre, Bryum pseudotriquetrum, Hypnum lindbergii* as well as *Thuidium delicatulum* and several Sphagnums.

In northern areas be sure to look for some rarities, such as *Meesia triquetra, Paludella squarrosa* and *Pseudocalliergon trifarium.*

Searching for rare bryos at Chickering Bog in Vermont

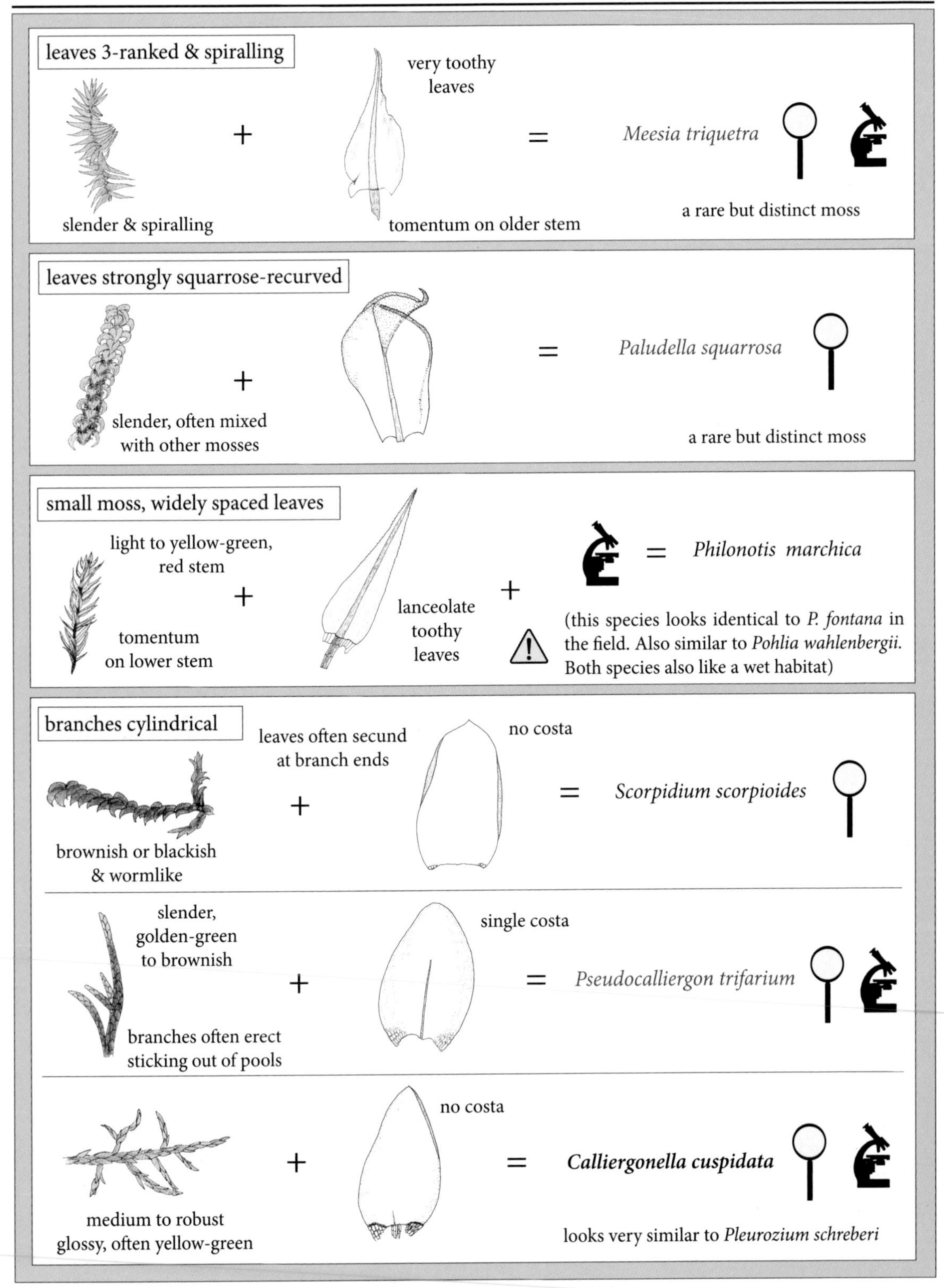
leaves 3-ranked & spiralling
very toothy leaves
+
=
Meesia triquetra
slender & spiralling
tomentum on older stem
a rare but distinct moss
leaves strongly squarrose-recurved
+
=
Paludella squarrosa
slender, often mixed with other mosses
a rare but distinct moss
small moss, widely spaced leaves
light to yellow-green, red stem
+
tomentum on lower stem
lanceolate toothy leaves
+
=
Philonotis marchica
(this species looks identical to P. fontana in the field. Also similar to Pohlia wahlenbergii. Both species also like a wet habitat)
branches cylindrical
leaves often secund at branch ends
no costa
+
=
Scorpidium scorpioides
brownish or blackish & wormlike
slender, golden-green to brownish
single costa
+
=
Pseudocalliergon trifarium
branches often erect sticking out of pools
no costa
+
=
Calliergonella cuspidata
medium to robust glossy, often yellow-green
looks very similar to Pleurozium schreberi

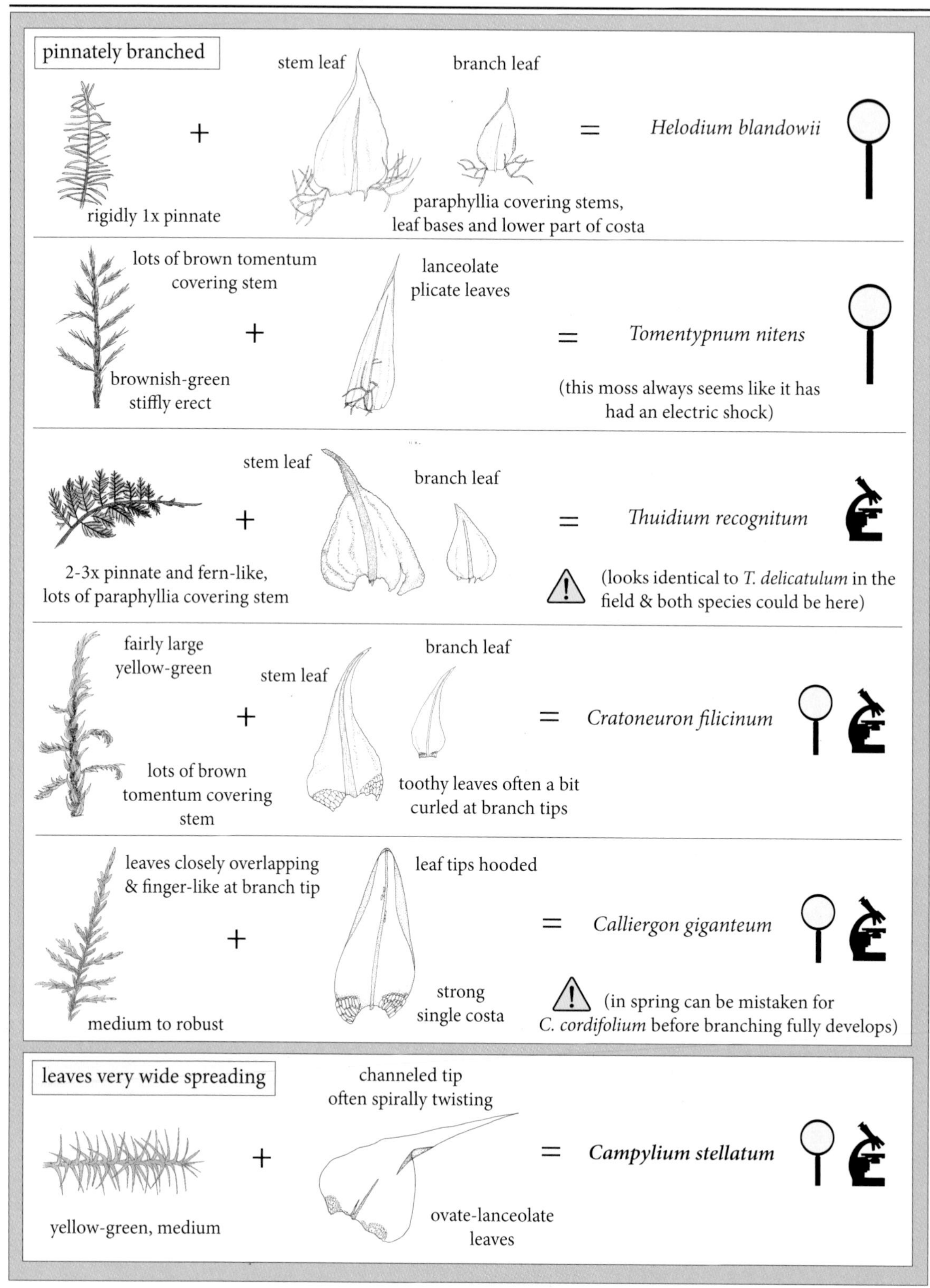
pinnately branched
stem leaf
branch leaf
+
=
Helodium blandowii
rigidly 1x pinnate
paraphyllia covering stems,
leaf bases and lower part of costa
lots of brown tomentum
covering stem
lanceolate
plicate leaves
+
=
Tomentypnum nitens
brownish-green
stiffly erect
(this moss always seems like it has
had an electric shock)
stem leaf
branch leaf
+
=
Thuidium recognitum
2-3x pinnate and fern-like,
lots of paraphyllia covering stem
(looks identical to *T. delicatulum* in the
field & both species could be here)
fairly large
yellow-green
stem leaf
branch leaf
+
=
Cratoneuron filicinum
lots of brown
tomentum covering
stem
toothy leaves often a bit
curled at branch tips
leaves closely overlapping
& finger-like at branch tip
leaf tips hooded
+
=
Calliergon giganteum
medium to robust
strong
single costa
(in spring can be mistaken for
C. cordifolium before branching fully develops)
leaves very wide spreading
channeled tip
often spirally twisting
+
=
Campylium stellatum
yellow-green, medium
ovate-lanceolate
leaves

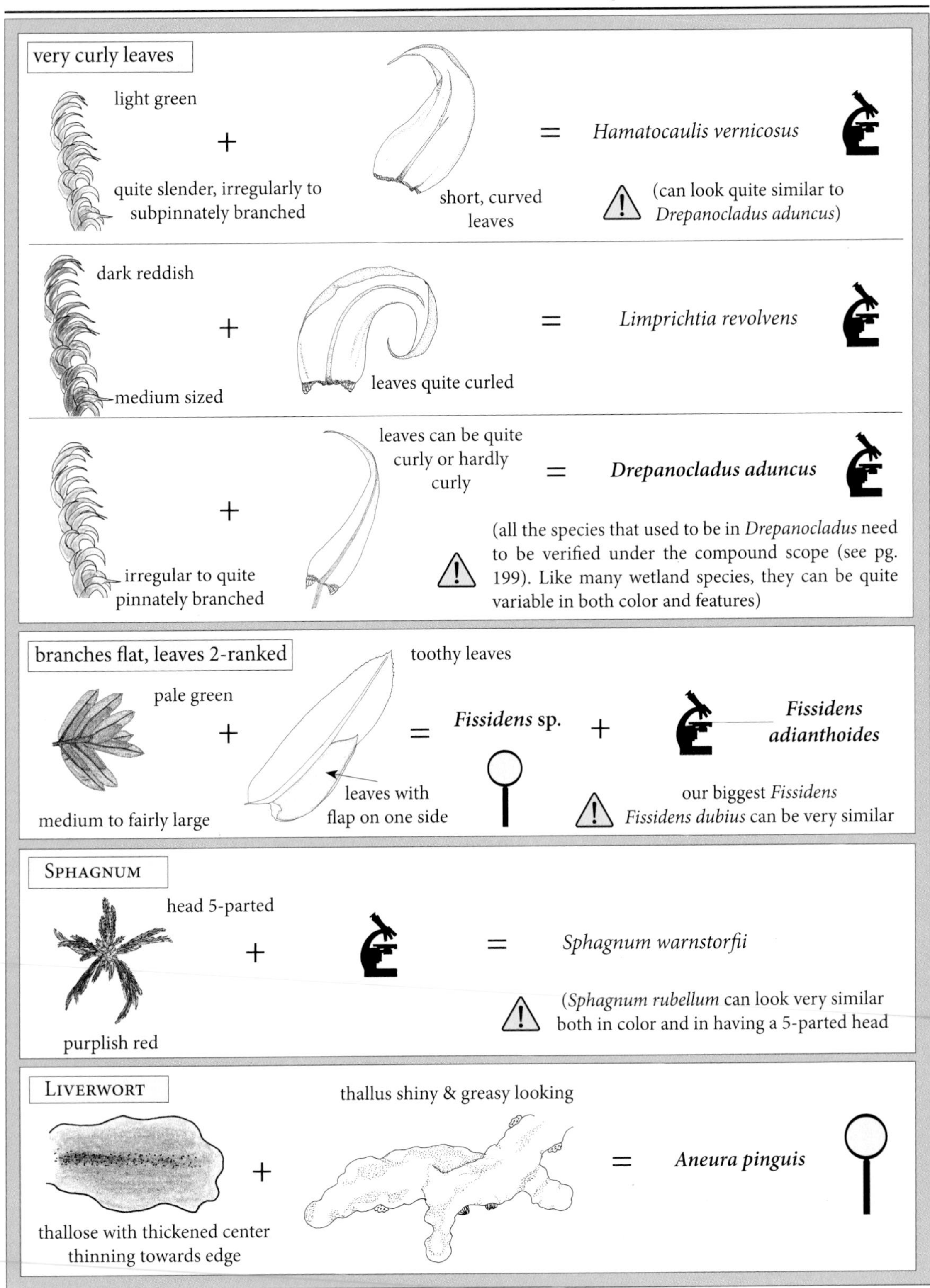
very curly leaves
light green
+
quite slender, irregularly to subpinnately branched
short, curved leaves
=
Hamatocaulis vernicosus
(can look quite similar to Drepanocladus aduncus)
dark reddish
+
medium sized
leaves quite curled
=
Limprichtia revolvens
leaves can be quite curly or hardly curly
+
irregular to quite pinnately branched
=
Drepanocladus aduncus
(all the species that used to be in Drepanocladus need to be verified under the compound scope (see pg. 199). Like many wetland species, they can be quite variable in both color and features)
branches flat, leaves 2-ranked
pale green
+
medium to fairly large
toothy leaves
leaves with flap on one side
=
Fissidens sp.
+
Fissidens adianthoides
our biggest Fissidens
Fissidens dubius can be very similar
Sphagnum
head 5-parted
+
purplish red
=
Sphagnum warnstorfii
(Sphagnum rubellum can look very similar both in color and in having a 5-parted head
Liverwort
thallose with thickened center thinning towards edge
+
thallus shiny & greasy looking
=
Aneura pinguis

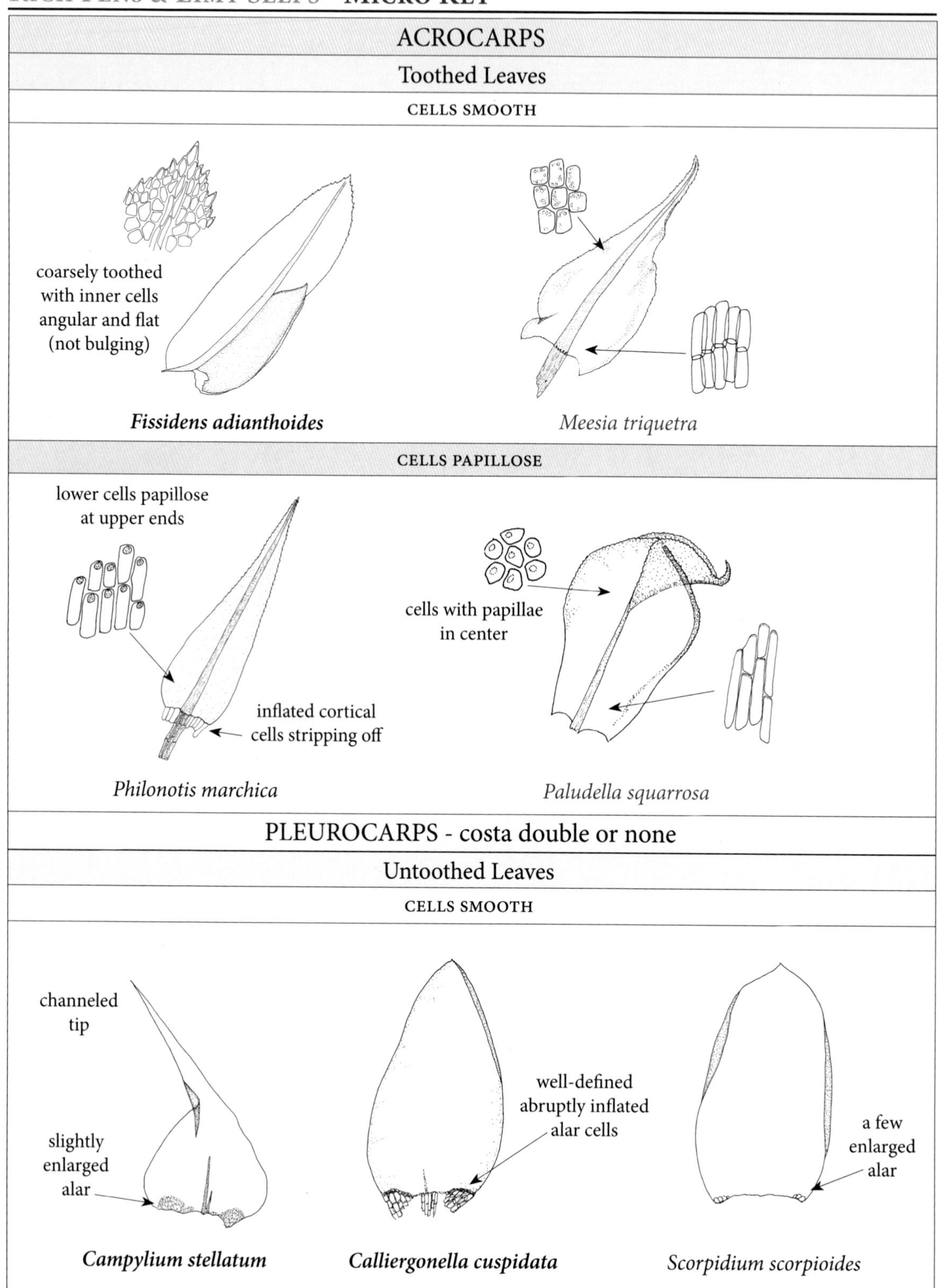
ACROCARPS
Toothed Leaves
CELLS SMOOTH
coarsely toothed
with inner cells
angular and flat
(not bulging)
Fissidens adianthoides
Meesia triquetra
CELLS PAPILLOSE
lower cells papillose
at upper ends
inflated cortical
cells stripping off
Philonotis marchica
cells with papillae
in center
Paludella squarrosa
PLEUROCARPS - costa double or none
Untoothed Leaves
CELLS SMOOTH
channeled
tip
slightly
enlarged
alar
Campylium stellatum
well-defined
abruptly inflated
alar cells
Calliergonella cuspidata
a few
enlarged
alar
Scorpidium scorpioides

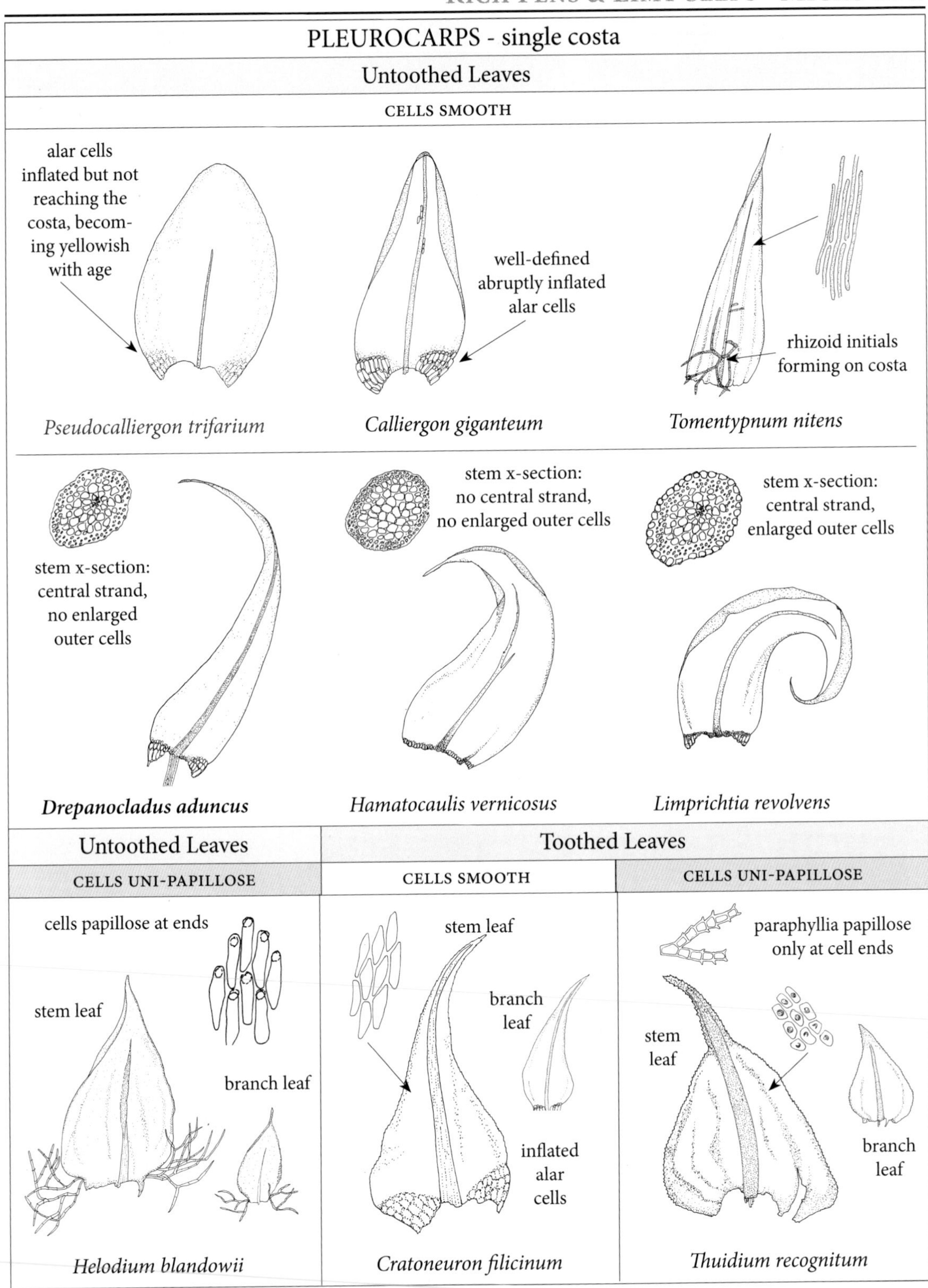
PLEUROCARPS - single costa
Untoothed Leaves
CELLS SMOOTH
alar cells inflated but not reaching the costa, becoming yellowish with age
Pseudocalliergon trifarium
well-defined abruptly inflated alar cells
Calliergon giganteum
rhizoid initials forming on costa
Tomentypnum nitens
stem x-section: central strand, no enlarged outer cells
Drepanocladus aduncus
stem x-section: no central strand, no enlarged outer cells
Hamatocaulis vernicosus
stem x-section: central strand, enlarged outer cells
Limprichtia revolvens
Untoothed Leaves
Toothed Leaves
CELLS UNI-PAPILLOSE
CELLS SMOOTH
CELLS UNI-PAPILLOSE
cells papillose at ends
stem leaf
branch leaf
Helodium blandowii
stem leaf
branch leaf
inflated alar cells
Cratoneuron filicinum
paraphyllia papillose only at cell ends
stem leaf
branch leaf
Thuidium recognitum

Meesia triquetra

Paludella squarrosa

Campylium stellatum

Calliergonella cuspidata

Scorpidium scorpioides

Drepanocladus aduncus

Thuidium recognitum

Tomentypnum nitens

Helodium blandowii

Calliergon giganteum

Sphagnum warnstorfii

Aneura pinguis

DUNG MOSSES

If you find yourself in northern country in boggy woods or high up in elevation, now is the time to look for dung mosses. Some of them you might actually mistake for a fungus, so unusual is their capsule shape! *Splachnum* also has an unusual feature in that its seta often continues to grow even after the capsule has opened. They all take advantage of flies to disperse their spores. Beautifully colored & often growing in masses, they are a wonder to see.

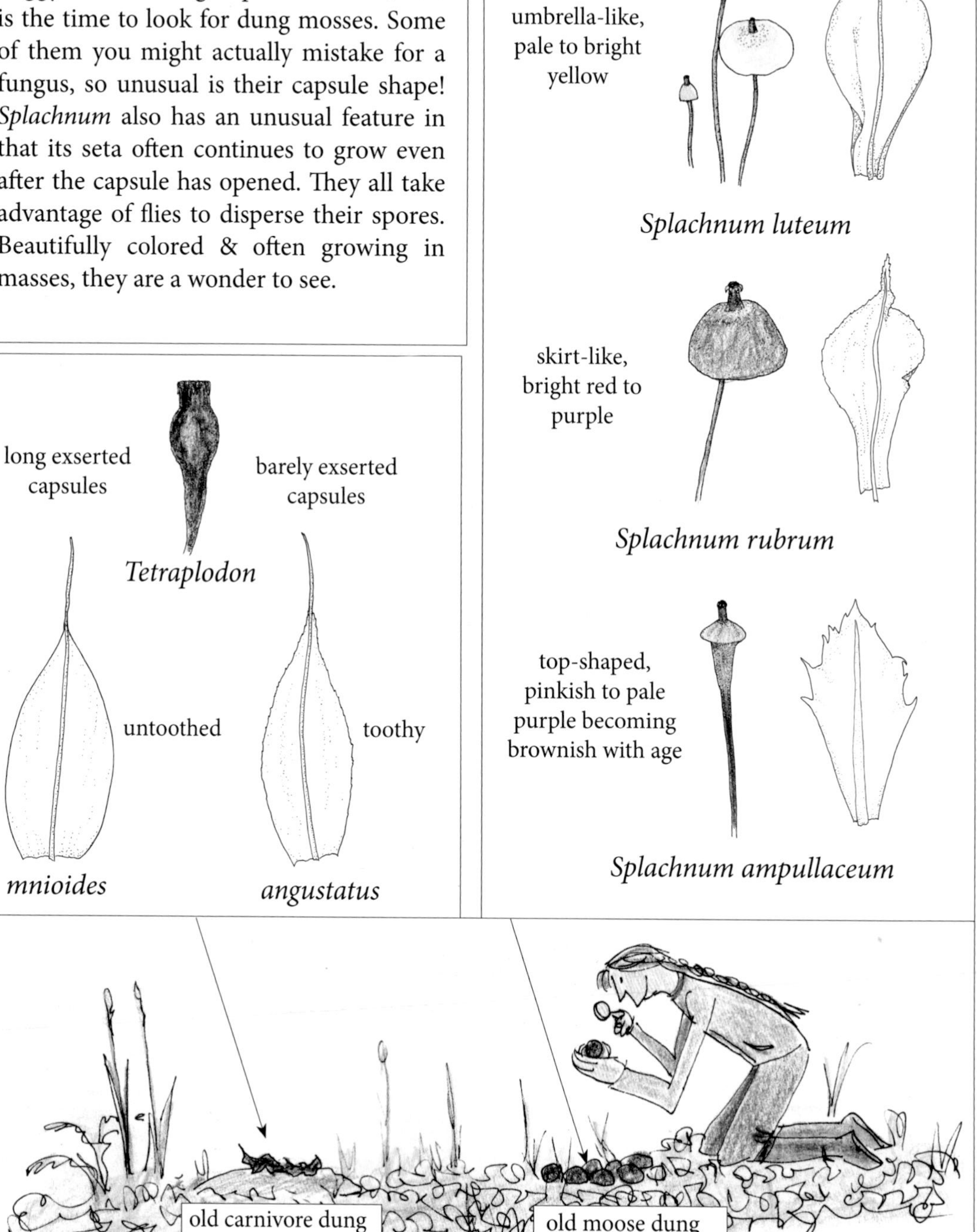

BRYOPHYTES
in the
REAL WORLD
Putting It All Together

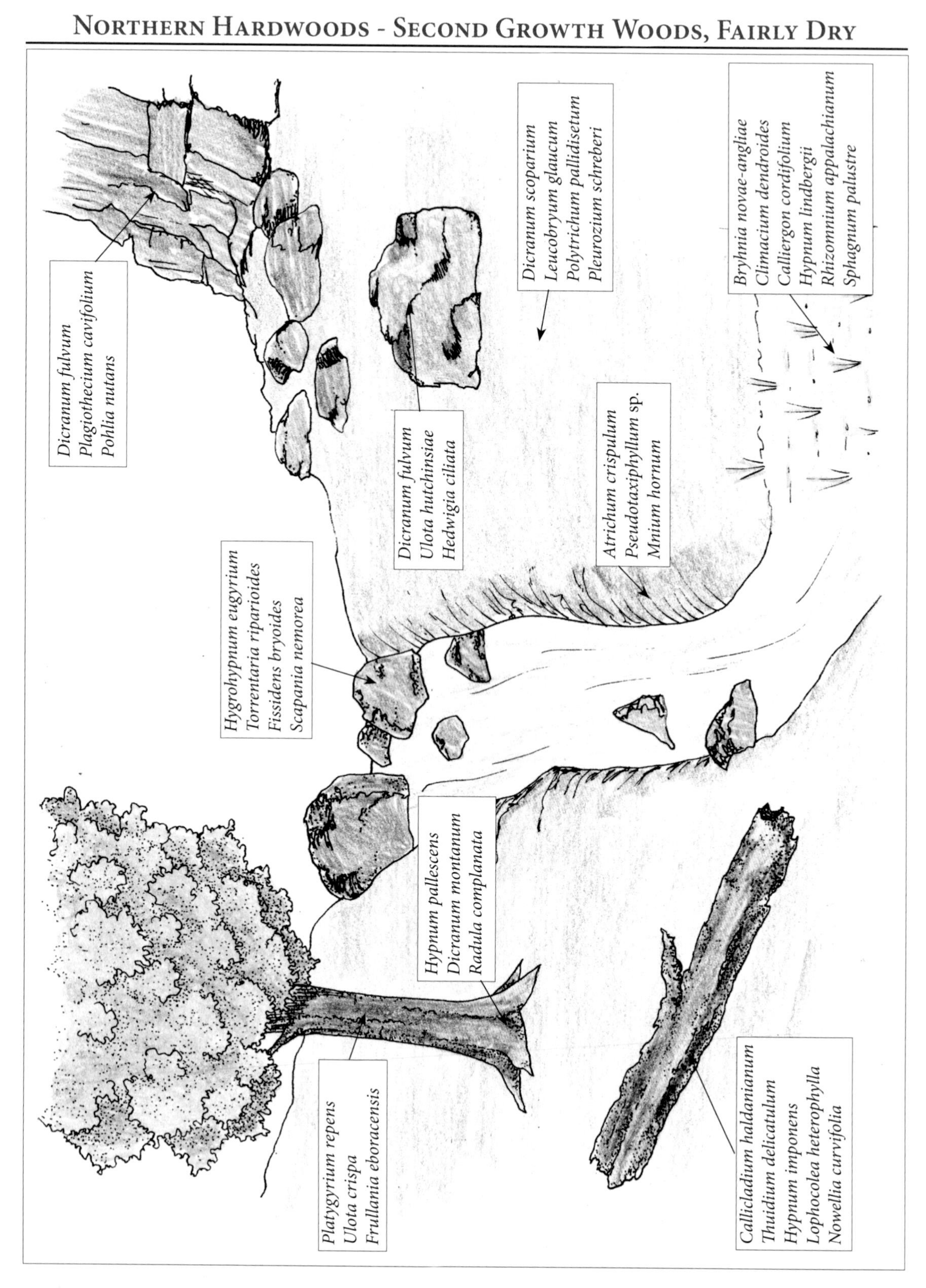
Dicranum fulvum
Plagiothecium cavifolium
Pohlia nutans
Dicranum scoparium
Leucobryum glaucum
Polytrichum pallidisetum
Pleurozium schreberi
Brynhia novae-angliae
Climacium dendroides
Calliergon cordifolium
Hypnum lindbergii
Rhizomnium appalachianum
Sphagnum palustre
Dicranum fulvum
Ulota hutchinsiae
Hedwigia ciliata
Atrichum crispulum
Pseudotaxiphyllum sp.
Mnium hornum
Hygrohypnum eugyrium
Torrentaria riparioides
Fissidens bryoides
Scapania nemorea
Hypnum pallescens
Dicranum montanum
Radula complanata
Platygyrium repens
Ulota crispa
Frullania eboracensis
Callicladium haldanianum
Thuidium delicatulum
Hypnum imponens
Lophocolea heterophylla
Nowellia curvifolia

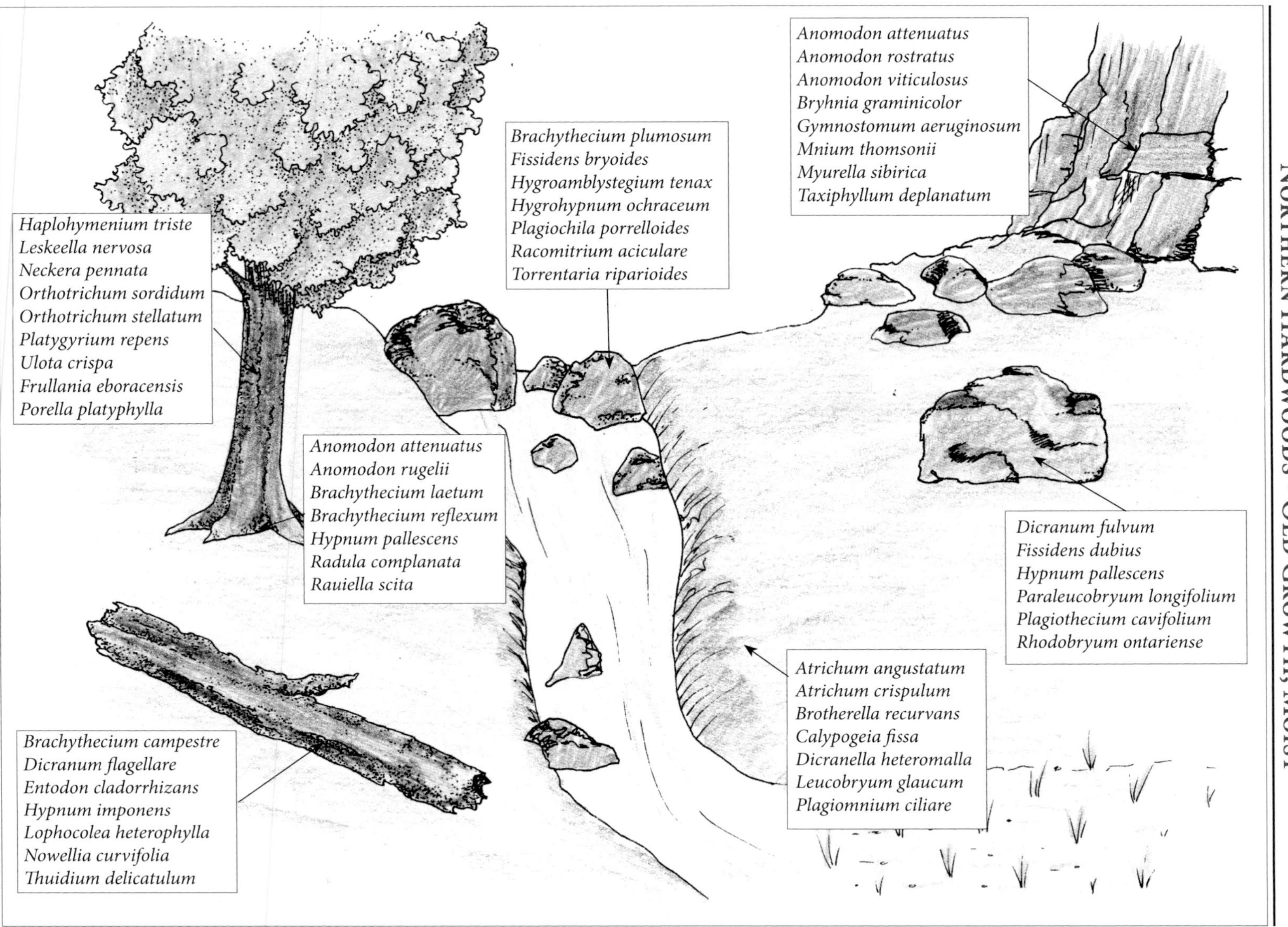

Haplohymenium triste
Leskeella nervosa
Neckera pennata
Orthotrichum sordidum
Orthotrichum stellatum
Platygyrium repens
Ulota crispa
Frullania eboracensis
Porella platyphylla

Brachythecium plumosum
Fissidens bryoides
Hygroamblystegium tenax
Hygrohypnum ochraceum
Plagiochila porrelloides
Racomitrium aciculare
Torrentaria riparioides

Anomodon attenuatus
Anomodon rostratus
Anomodon viticulosus
Bryhnia graminicolor
Gymnostomum aeruginosum
Mnium thomsonii
Myurella sibirica
Taxiphyllum deplanatum

Anomodon attenuatus
Anomodon rugelii
Brachythecium laetum
Brachythecium reflexum
Hypnum pallescens
Radula complanata
Rauiella scita

Dicranum fulvum
Fissidens dubius
Hypnum pallescens
Paraleucobryum longifolium
Plagiothecium cavifolium
Rhodobryum ontariense

Atrichum angustatum
Atrichum crispulum
Brotherella recurvans
Calypogeia fissa
Dicranella heteromalla
Leucobryum glaucum
Plagiomnium ciliare

Brachythecium campestre
Dicranum flagellare
Entodon cladorrhizans
Hypnum imponens
Lophocolea heterophylla
Nowellia curvifolia
Thuidium delicatulum

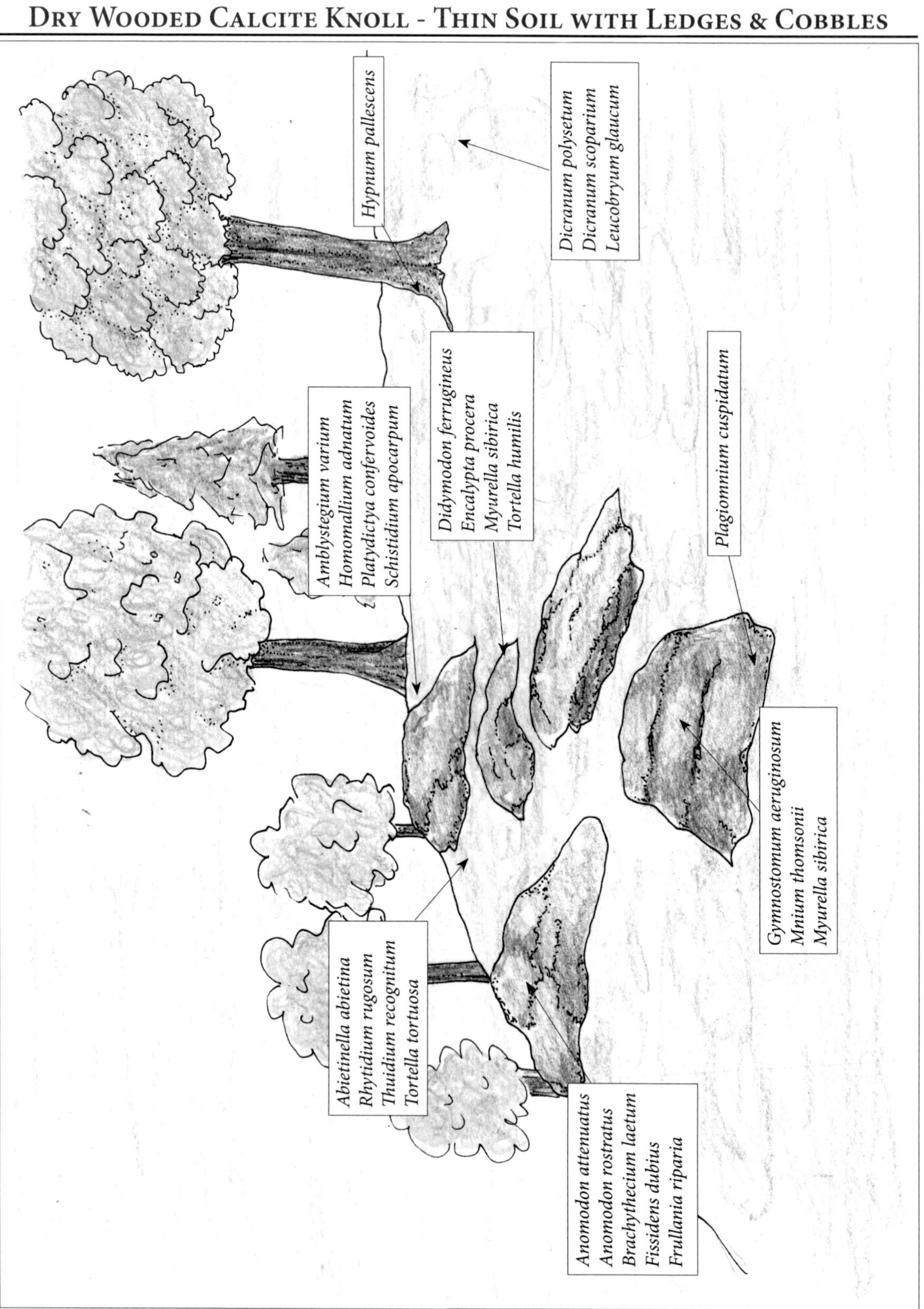
Hypnum pallescens
Dicranum polysetum
Dicranum scoparium
Leucobryum glaucum
Amblystegium varium
Homomallium adnatum
Platydictya confervoides
Schistidium apocarpum
Didymodon ferrugineus
Encalypta procera
Myurella sibirica
Tortella humilis
Plagiomnium cuspidatum
Abietinella abietina
Rhytidium rugosum
Thuidium recognitum
Tortella tortuosa
Gymnostomum aeruginosum
Mnium thomsonii
Myurella sibirica
Anomodon attenuatus
Anomodon rostratus
Brachythecium laetum
Fissidens dubius
Frullania riparia

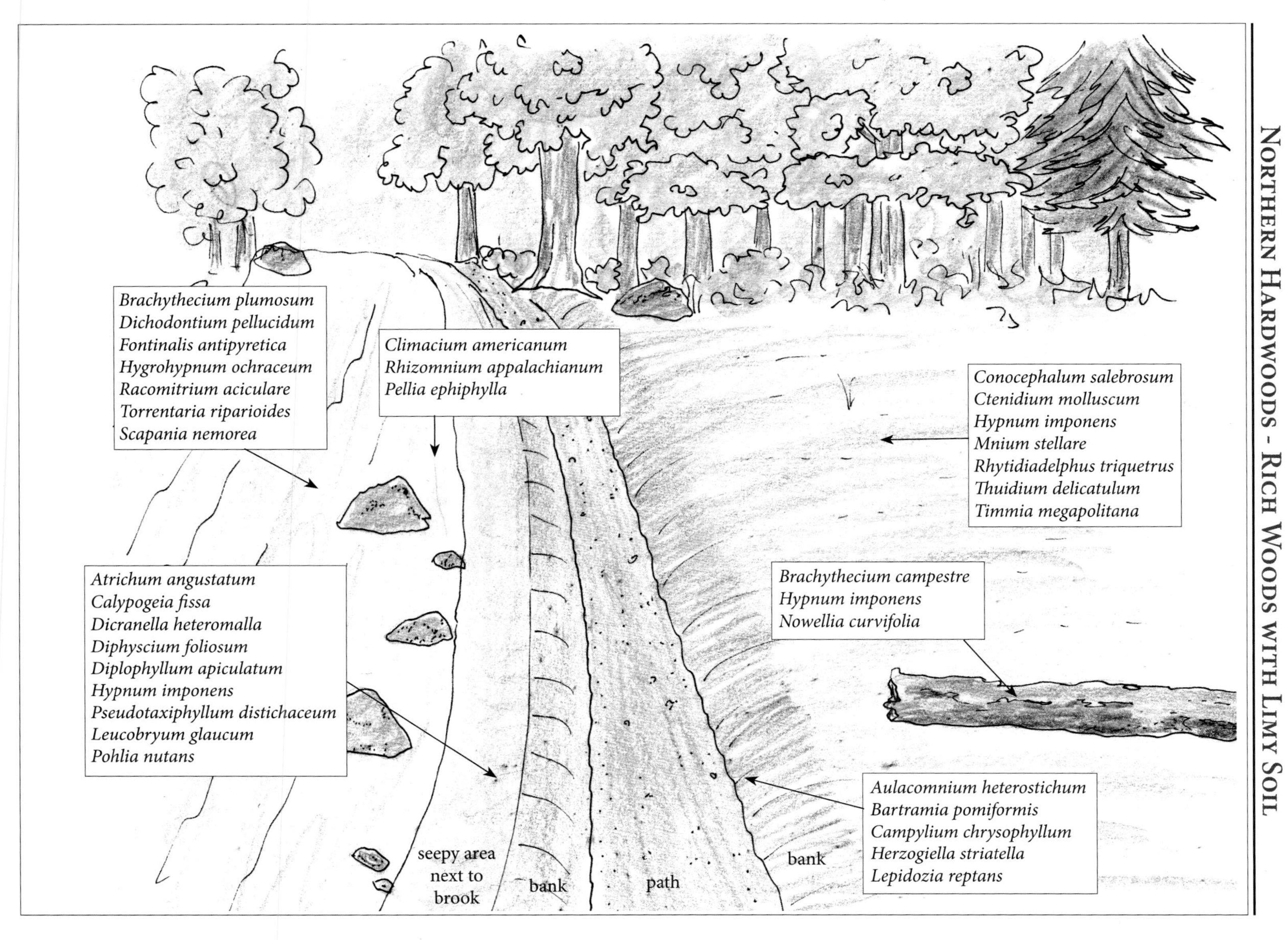
Brachythecium plumosum
Dichodontium pellucidum
Fontinalis antipyretica
Hygrohypnum ochraceum
Racomitrium aciculare
Torrentaria riparioides
Scapania nemorea
Climacium americanum
Rhizomnium appalachianum
Pellia ephiphylla
Conocephalum salebrosum
Ctenidium molluscum
Hypnum imponens
Mnium stellare
Rhytidiadelphus triquetrus
Thuidium delicatulum
Timmia megapolitana
Atrichum angustatum
Calypogeia fissa
Dicranella heteromalla
Diphyscium foliosum
Diplophyllum apiculatum
Hypnum imponens
Pseudotaxiphyllum distichaceum
Leucobryum glaucum
Pohlia nutans
Brachythecium campestre
Hypnum imponens
Nowellia curvifolia
Aulacomnium heterostichum
Bartramia pomiformis
Campylium chrysophyllum
Herzogiella striatella
Lepidozia reptans
seepy area
next to
brook
bank
path
bank

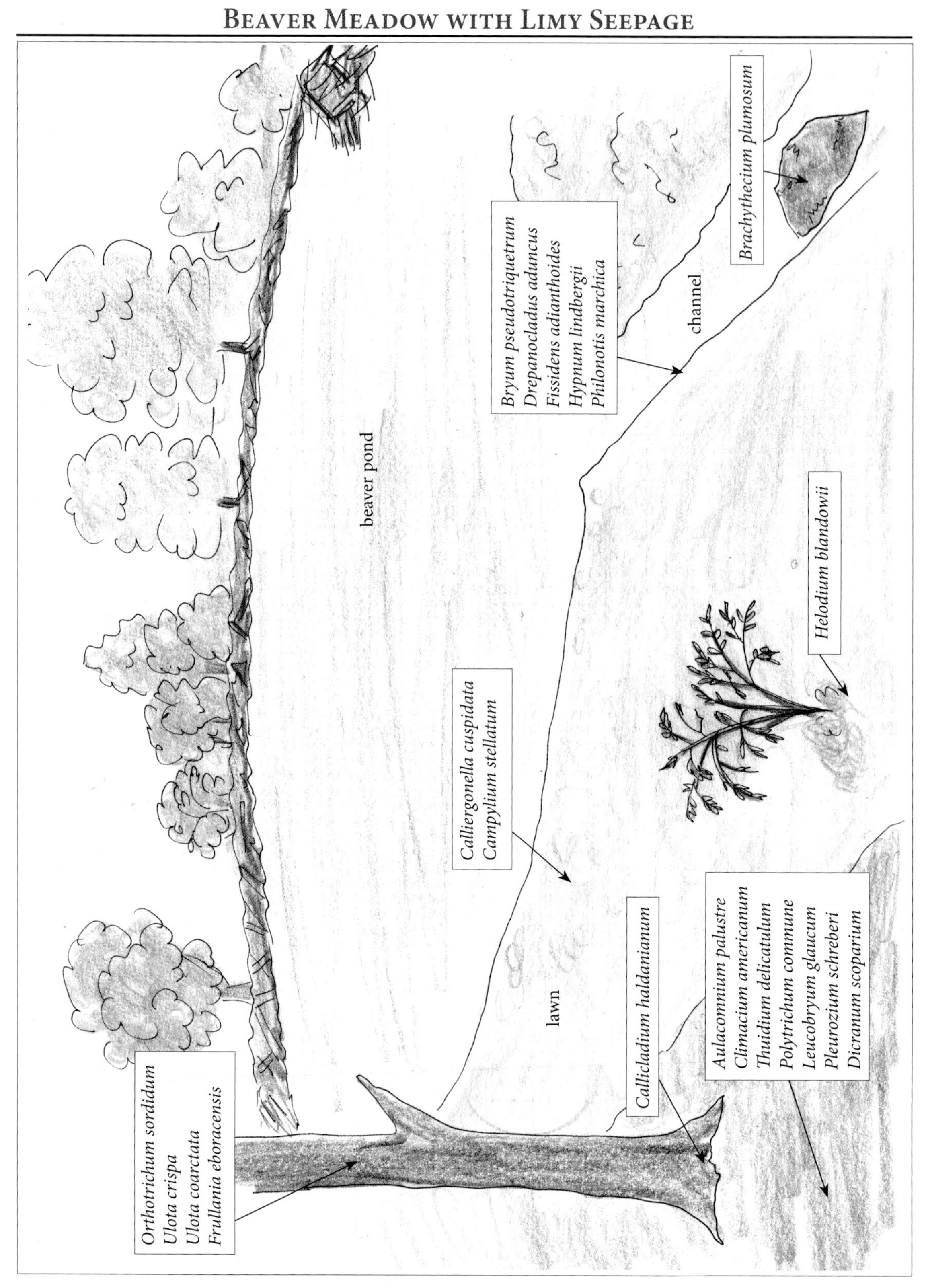

Orthotrichum sordidum
Ulota crispa
Ulota coarctata
Frullania eboracensis
lawn
Callicladium haldanianum
Aulacomnium palustre
Climacium americanum
Thuidium delicatulum
Polytrichum commune
Leucobryum glaucum
Pleurozium schreberi
Dicranum scoparium
Calliergonella cuspidata
Campylium stellatum
beaver pond
Helodium blandowii
Bryum pseudotriquetrum
Drepanocladus aduncus
Fissidens adianthoides
Hypnum lindbergii
Philonotis marchica
channel
Brachythecium plumosum

Bazzania trilobata
Brotherella recurvans
Cephalozia lunulifolia
Dicranum scoparium
Hypnum imponens
Lepidozia reptans
Tetraphis pellucida
A bog mat:
Sphagnum magellanicum
Sphagnum fuscum
Sphagnum fallax
Sphagnum centrale
Sphagnum cuspidatum
Bazzania trilobata
Cephalozia lunulifolia
Dicranum scoparium
Mylia anomala
Pleurozium schreberi
Polytrichum strictum
Ptilidium ciliare
Dicranum montanum
Hypnum pallescens
Ptilidium pulcherrimum
Warnstorfia fluitans
Sphagnum palustre
Sphagnum capillifolium
Callicladium haldanianum
Calypogeia sphagnicola
Dicranum flagellare
Plagiothecium laetum
Thuidium delicatulum
moat

Platygyrium repens
Ulota crispa
Frullania eboracensis
Brotherella recurvans
Hypnum imponens
Hypnum pallescens
Tetraphis pellucida
Atrichum crispulum
Bazzania trilobata
Dicranum scoparium
Lepidozia reptans
Polytrichum pallidisetum
Pseudotaxiphyllum distichaceum
Callicladium haldanianum
Dicranum flagellare
Oncophorus wahlenbergii
Ptilidium pulcherrimum
Rhizomnium punctatum
Thuidium delicatulum
hummocks
Aulacomnium palustre
Bryhnia novae-angliae
Pallavicinia lyellii
Rhytidiadelphus squarrosus
Sphagnum girgensohnii
Sphagnum palustre
Sphagnum squarrosus
pools
Calliergon cordifolium
Rhizomnium appalachianum

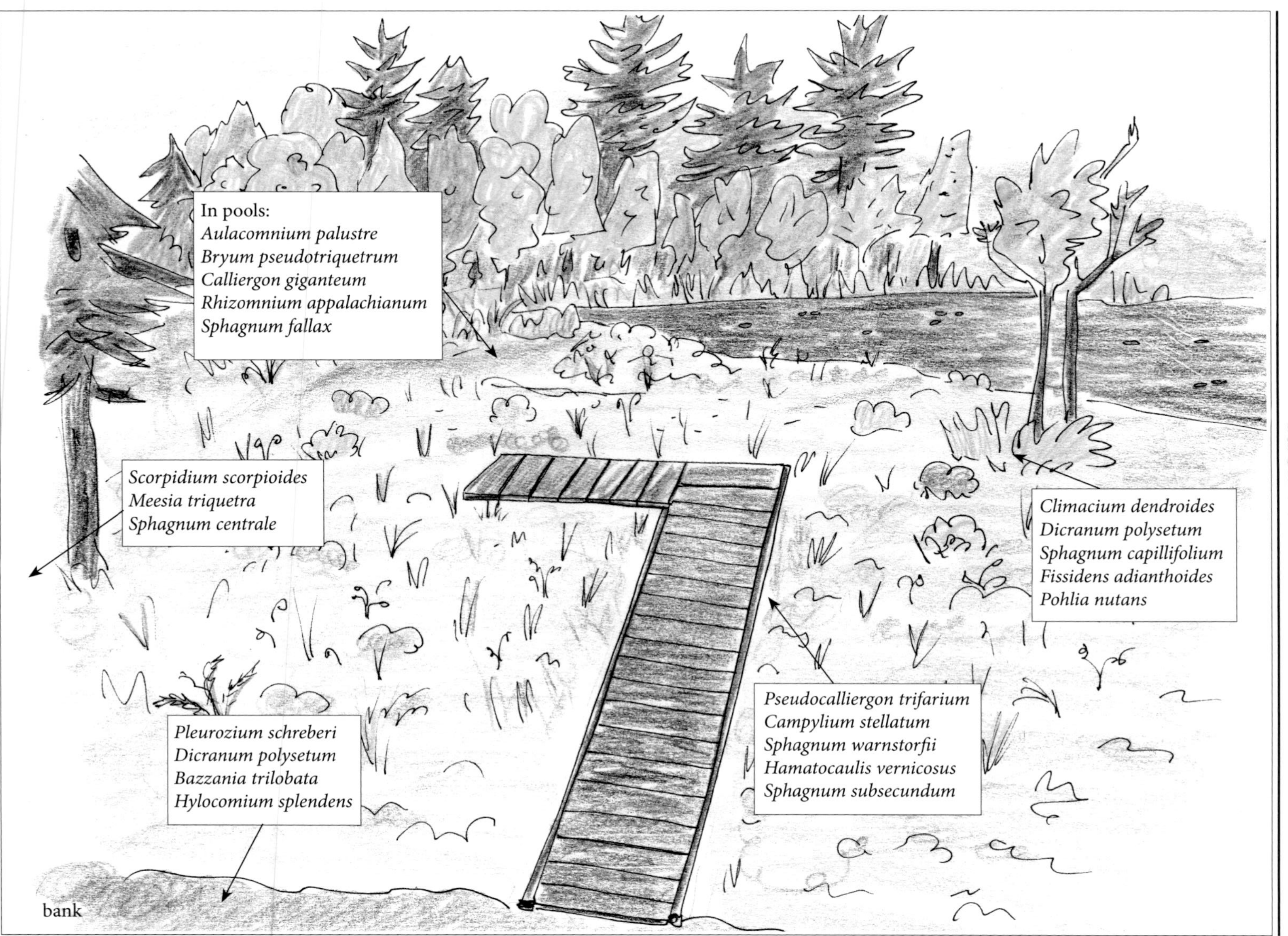
In pools:
Aulacomnium palustre
Bryum pseudotriquetrum
Calliergon giganteum
Rhizomnium appalachianum
Sphagnum fallax
Scorpidium scorpioides
Meesia triquetra
Sphagnum centrale
Climacium dendroides
Dicranum polysetum
Sphagnum capillifolium
Fissidens adianthoides
Pohlia nutans
Pseudocalliergon trifarium
Campylium stellatum
Sphagnum warnstorfii
Hamatocaulis vernicosus
Sphagnum subsecundum
Pleurozium schreberi
Dicranum polysetum
Bazzania trilobata
Hylocomium splendens
bank

Powerline Cut - Disturbed Soil with Wet Spot

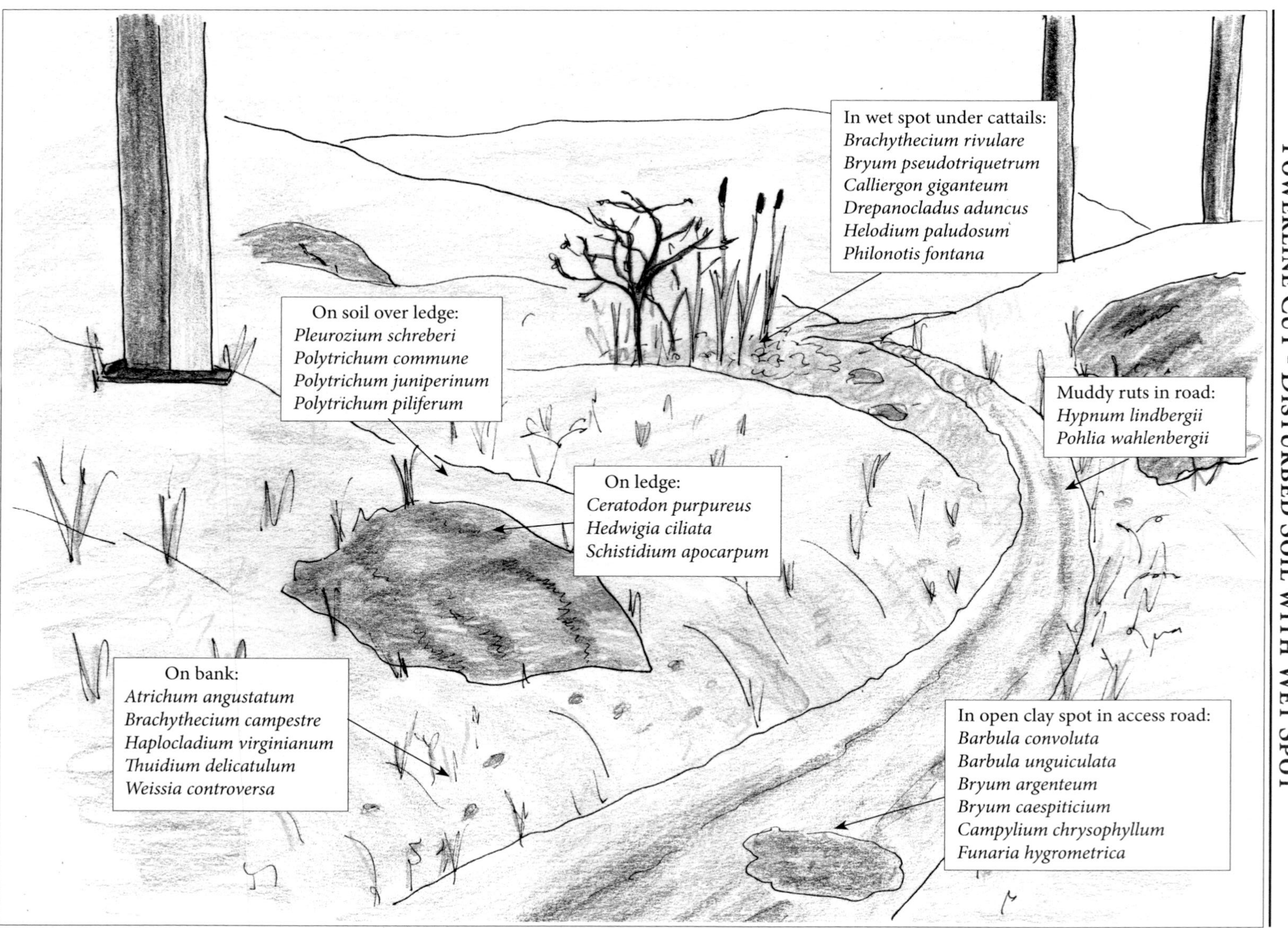

A Closer Look at Some of the Bigger

MOSS GROUPS

The big moss groups...now you are in dangerous territory! In the beginning you might want to just skip this part as it can get quite confusing, and you often need a compound microscope for some of the finer details.

It might take some time for you to develop the confidence and skills to be able to make acceptable cross sections of leaves, stems and capsules, which often are required when determining some of the species. Also, time spent looking at many specimens to see the variability within each species is helpful in understanding when something is a different species or just a variation within a species.

It was many years before I had the courage or know-how to delve into the realm of *Orthotrichum* and even later into *Sphagnum*. It is also OK if you never want to get to this point...if you're happy with just going out into the field and saying, "Oh, that's a *Brachythecium*" or "That's *Fissidens.*" That is an accomplishment in itself!

I won't delude you into thinking that these are easy groups. They aren't. As with field characters, there is often overlap in characters or some are missing. More often than not, nothing gets figured out, and you throw up your hands in frustration. We've all been there. Keep at it, keep looking, draw them (that's what helped me the most), and you will be able to figure them out.

Table of Contents to the Moss Groups

Atrichum.. 184
Polytrichum & Polytrichastrum.......... 185
Common Bryums................................ 186
Common Pohlias................................. 187
Mnium Group...................................... 188
Dicranum... 190
Dicranella & Ditrichum.................... 192
Orthotrichum.................................... 193
Fissidens... 194
Brachythecium.................................... 196
Amblystegiums.................................... 198
The Old Drepanocladus Group.......... 199
Hygrohypnums.................................... 200
Anomodons... 201

Atrichum

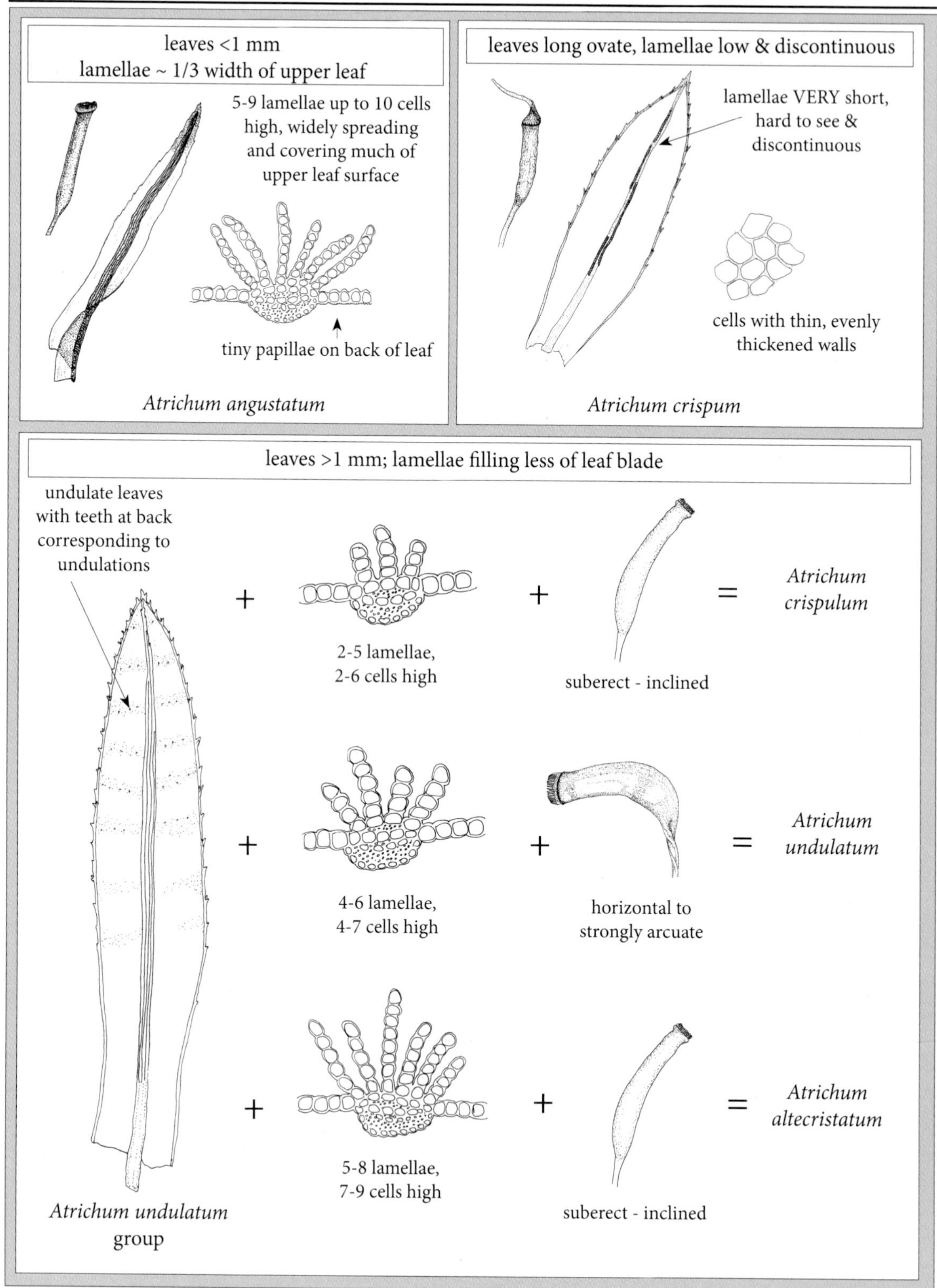

leaves <1 mm
lamellae ~ 1/3 width of upper leaf
5-9 lamellae up to 10 cells high, widely spreading and covering much of upper leaf surface
tiny papillae on back of leaf
Atrichum angustatum
leaves long ovate, lamellae low & discontinuous
lamellae VERY short, hard to see & discontinuous
cells with thin, evenly thickened walls
Atrichum crispum
leaves >1 mm; lamellae filling less of leaf blade
undulate leaves with teeth at back corresponding to undulations
+
2-5 lamellae, 2-6 cells high
+
suberect - inclined
=
Atrichum crispulum
+
4-6 lamellae, 4-7 cells high
+
horizontal to strongly arcuate
=
Atrichum undulatum
+
5-8 lamellae, 7-9 cells high
+
suberect - inclined
=
Atrichum altecristatum
Atrichum undulatum group

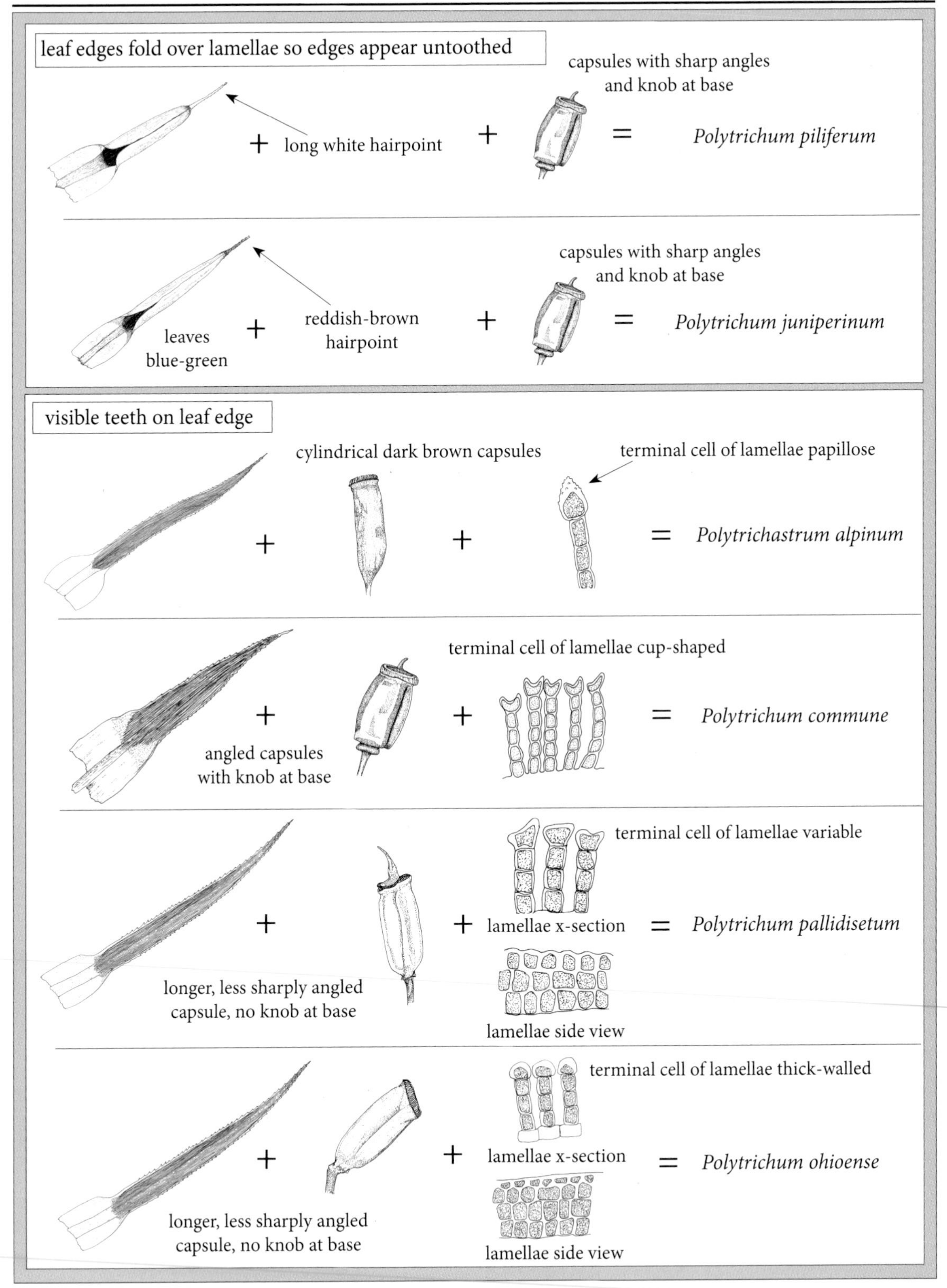
leaf edges fold over lamellae so edges appear untoothed
capsules with sharp angles and knob at base
+ long white hairpoint
+
=
Polytrichum piliferum
capsules with sharp angles and knob at base
leaves blue-green
+
reddish-brown hairpoint
+
=
Polytrichum juniperinum
visible teeth on leaf edge
cylindrical dark brown capsules
terminal cell of lamellae papillose
+
+
=
Polytrichastrum alpinum
terminal cell of lamellae cup-shaped
+
angled capsules with knob at base
+
=
Polytrichum commune
terminal cell of lamellae variable
+
+ lamellae x-section
=
Polytrichum pallidisetum
longer, less sharply angled capsule, no knob at base
lamellae side view
terminal cell of lamellae thick-walled
+
+
lamellae x-section
=
Polytrichum ohioense
longer, less sharply angled capsule, no knob at base
lamellae side view

Common Bryums

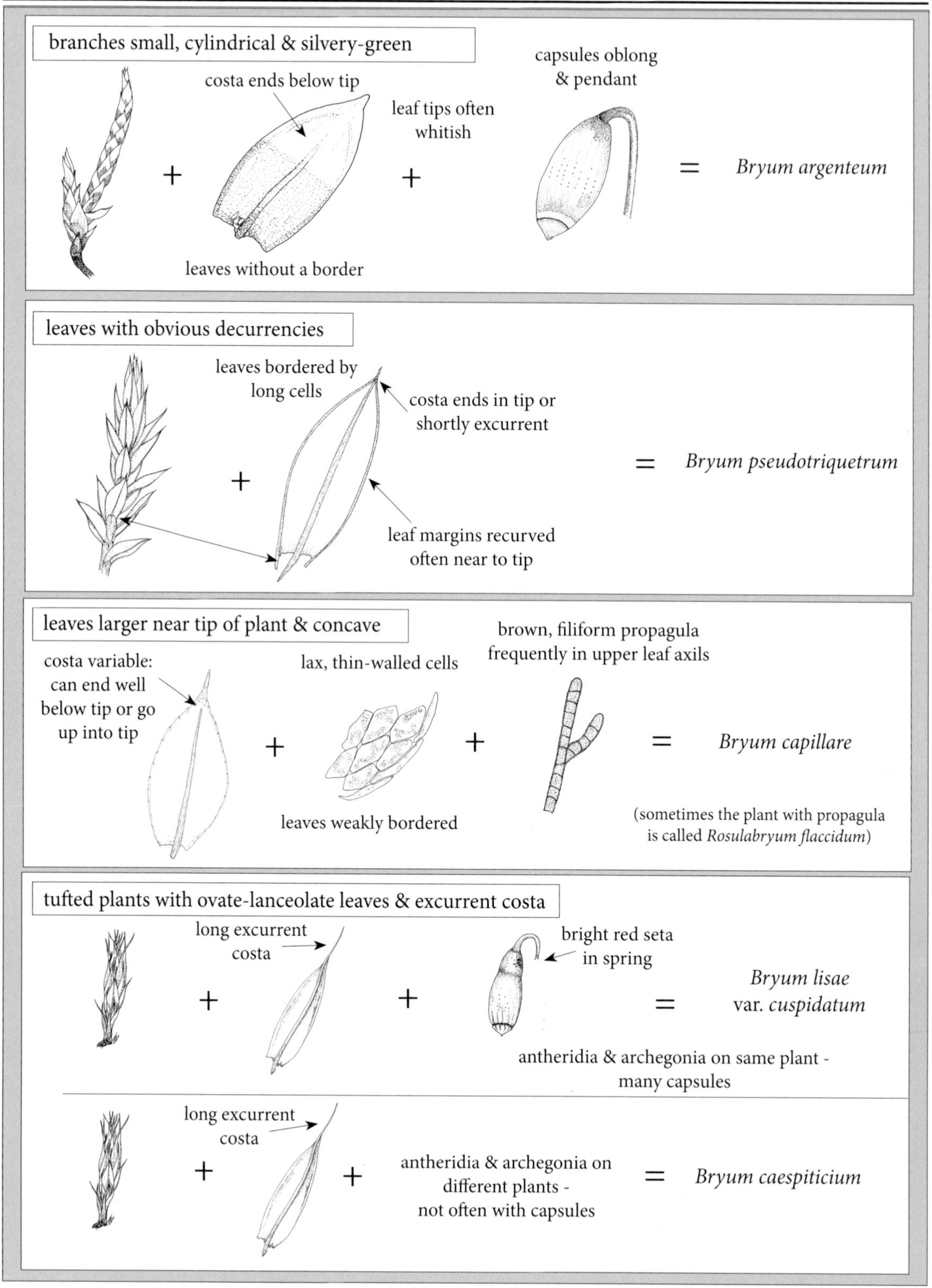

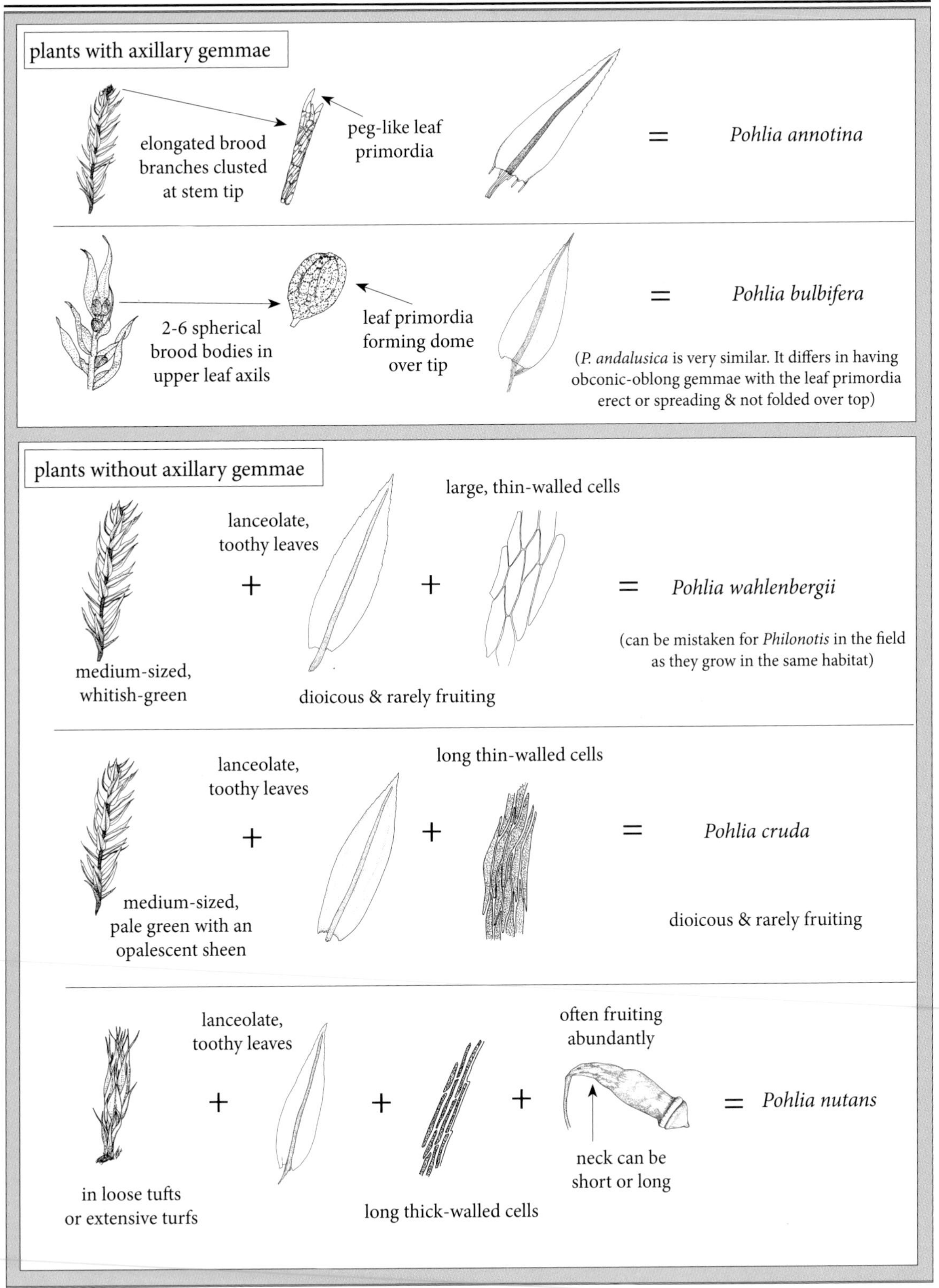
plants with axillary gemmae
elongated brood branches clusted at stem tip
peg-like leaf primordia
=
Pohlia annotina
2-6 spherical brood bodies in upper leaf axils
leaf primordia forming dome over tip
=
Pohlia bulbifera
(P. andalusica is very similar. It differs in having obconic-oblong gemmae with the leaf primordia erect or spreading & not folded over top)
plants without axillary gemmae
lanceolate, toothy leaves
+
large, thin-walled cells
+
=
Pohlia wahlenbergii
(can be mistaken for Philonotis in the field as they grow in the same habitat)
medium-sized, whitish-green
dioicous & rarely fruiting
lanceolate, toothy leaves
long thin-walled cells
+
+
=
Pohlia cruda
medium-sized, pale green with an opalescent sheen
dioicous & rarely fruiting
lanceolate, toothy leaves
often fruiting abundantly
+
+
+
=
Pohlia nutans
neck can be short or long
in loose tufts or extensive turfs
long thick-walled cells

MNIUM GROUP

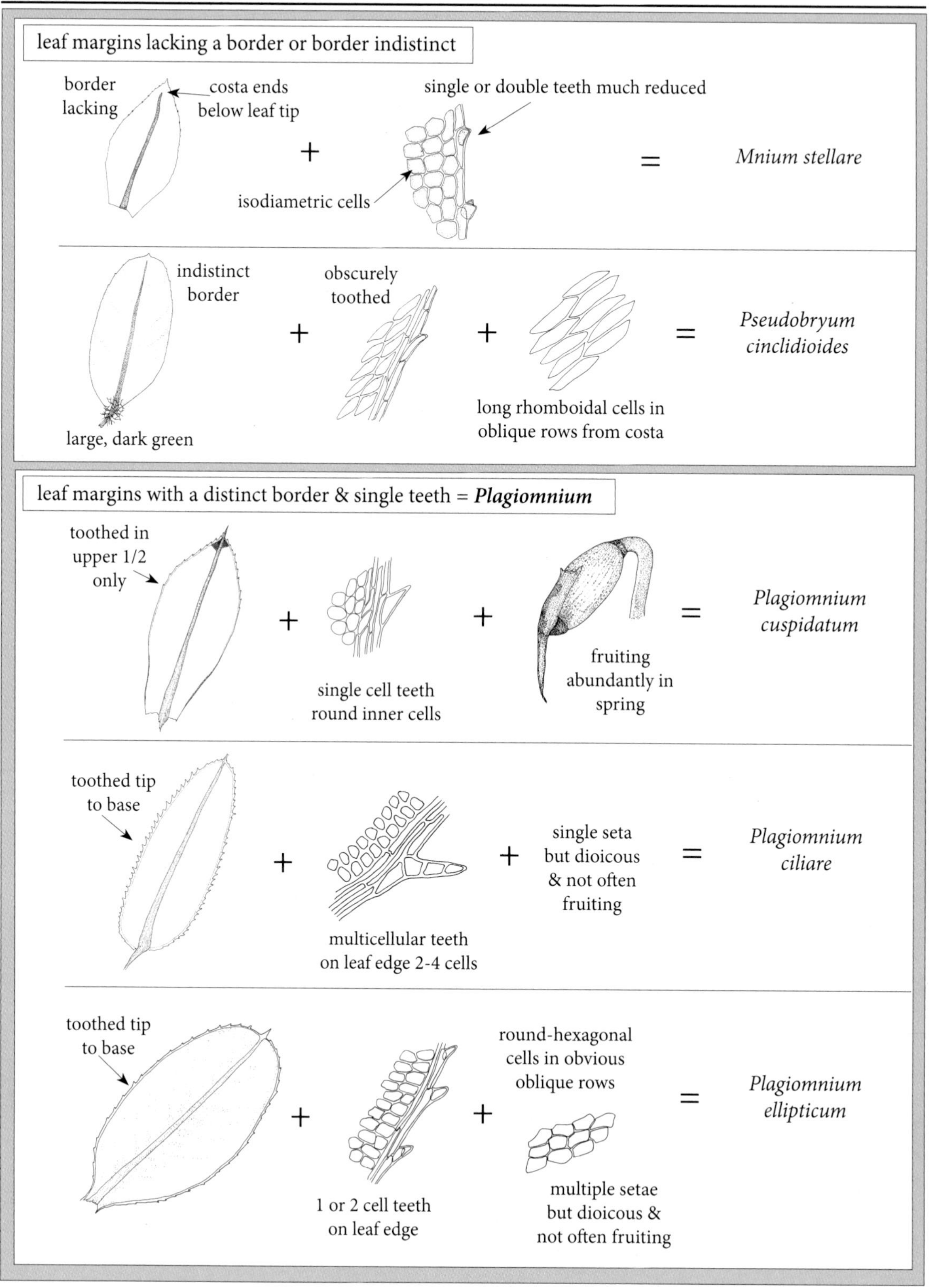

leaf margins lacking a border or border indistinct
border lacking
costa ends below leaf tip
+
single or double teeth much reduced
isodiametric cells
=
Mnium stellare
indistinct border
large, dark green
+
obscurely toothed
+
long rhomboidal cells in oblique rows from costa
=
Pseudobryum cinclidioides
leaf margins with a distinct border & single teeth = Plagiomnium
toothed in upper 1/2 only
+
single cell teeth round inner cells
+
fruiting abundantly in spring
=
Plagiomnium cuspidatum
toothed tip to base
+
multicellular teeth on leaf edge 2-4 cells
+
single seta but dioicous & not often fruiting
=
Plagiomnium ciliare
toothed tip to base
+
1 or 2 cell teeth on leaf edge
+
round-hexagonal cells in obvious oblique rows
multiple setae but dioicous & not often fruiting
=
Plagiomnium ellipticum

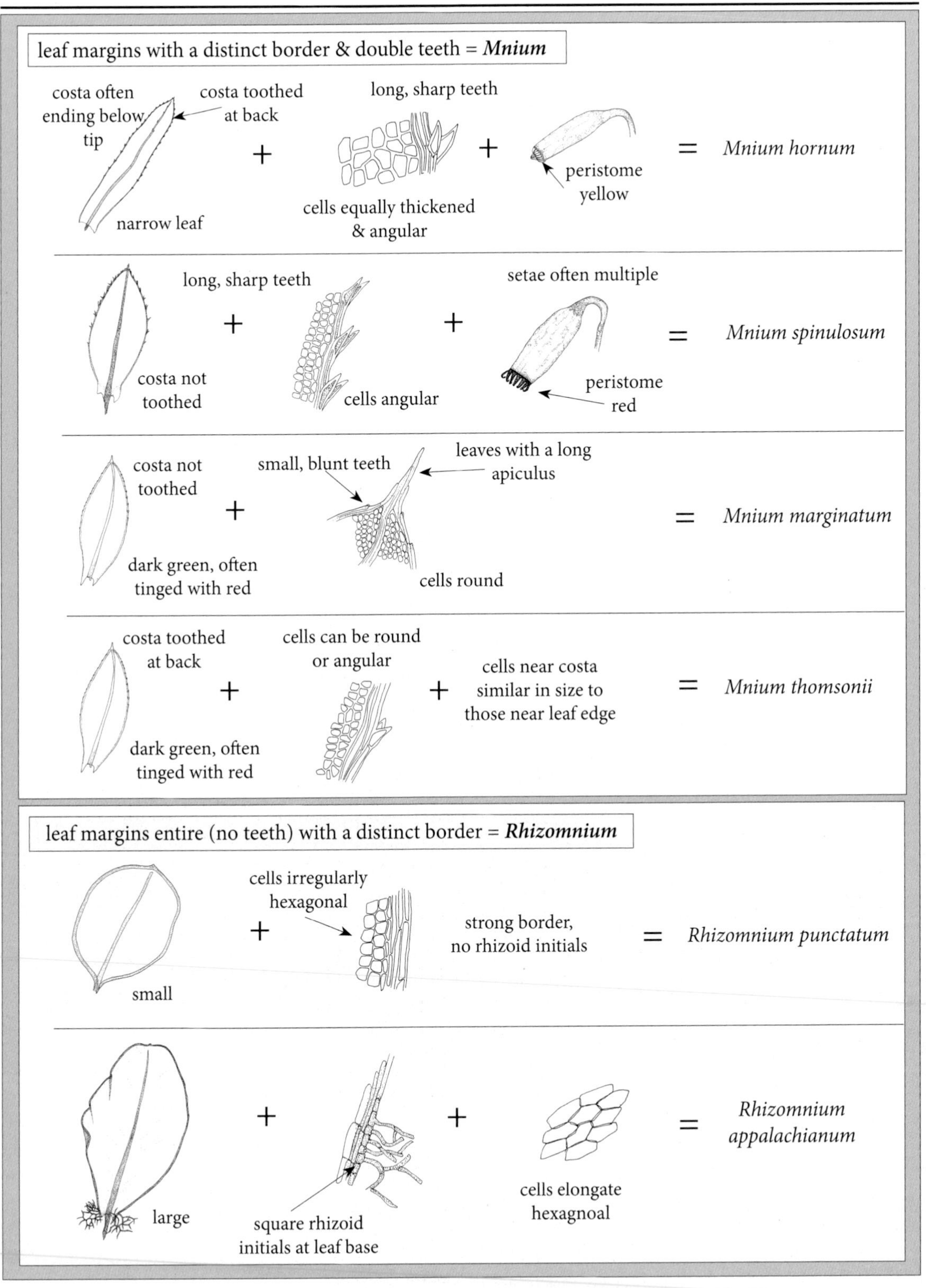
leaf margins with a distinct border & double teeth = Mnium
costa often ending below tip
costa toothed at back
narrow leaf
+
long, sharp teeth
cells equally thickened & angular
+
peristome yellow
= Mnium hornum
long, sharp teeth
+
cells angular
costa not toothed
+
setae often multiple
peristome red
= Mnium spinulosum
costa not toothed
+
small, blunt teeth
leaves with a long apiculus
dark green, often tinged with red
cells round
= Mnium marginatum
costa toothed at back
+
cells can be round or angular
+
cells near costa similar in size to those near leaf edge
dark green, often tinged with red
= Mnium thomsonii
leaf margins entire (no teeth) with a distinct border = Rhizomnium
small
+
cells irregularly hexagonal
strong border, no rhizoid initials
= Rhizomnium punctatum
large
+
square rhizoid initials at leaf base
+
cells elongate hexagnoal
= Rhizomnium appalachianum

DICRANUM

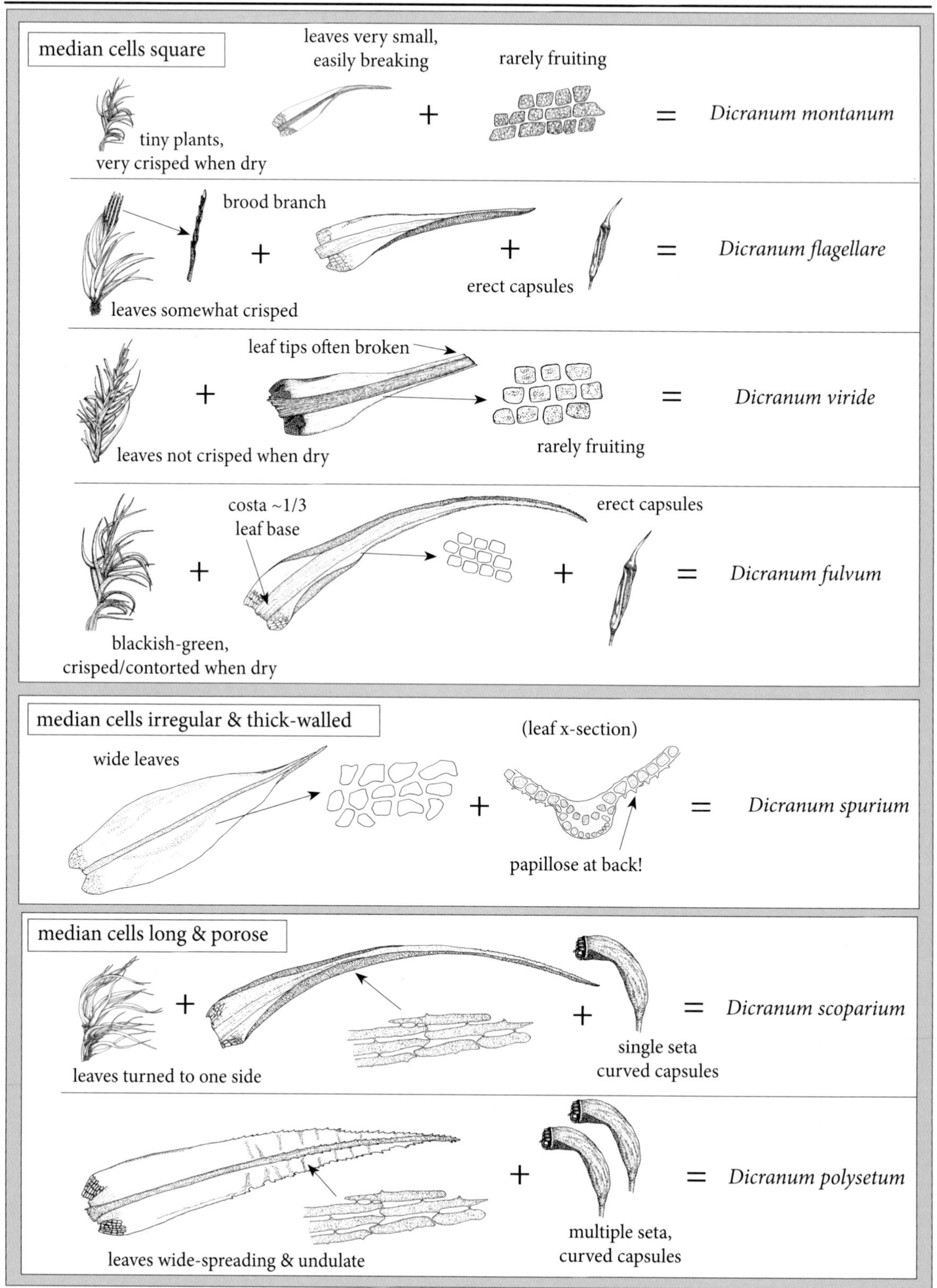

median cells square
leaves very small, easily breaking
rarely fruiting
tiny plants, very crisped when dry
+
=
Dicranum montanum
brood branch
+
+
erect capsules
=
Dicranum flagellare
leaves somewhat crisped
leaf tips often broken
+
=
Dicranum viride
leaves not crisped when dry
rarely fruiting
costa ~1/3 leaf base
erect capsules
+
+
=
Dicranum fulvum
blackish-green, crisped/contorted when dry
median cells irregular & thick-walled
(leaf x-section)
wide leaves
+
=
Dicranum spurium
papillose at back!
median cells long & porose
+
+
=
Dicranum scoparium
single seta curved capsules
leaves turned to one side
+
=
Dicranum polysetum
multiple seta, curved capsules
leaves wide-spreading & undulate

D. montanum & *D. viride* both have very fragile leaves that break easily and can propagate new plants. In *D. montanum*, the whole leaf breaks off, while in *D. viride* just the leaf tips break. *D. flagellare* has many brood branches clustered at the tip of the plant that easily become detached and grow into new plants.

D. scoparium and *D. polysetum* don't have any specialized structures or fragile parts, but they have a very interesting sexual reproductive strategy. The spores develop without any gender...if the spore lands on a suitable spot for growing, it will develop into a female plant. If, however, the spore lands on a nearby female plant, it will develop into a tiny dwarf male plant. A sensible arrangement as it increases the chances of fertilization due to a much closer proximity to the female.

FUN FACT!

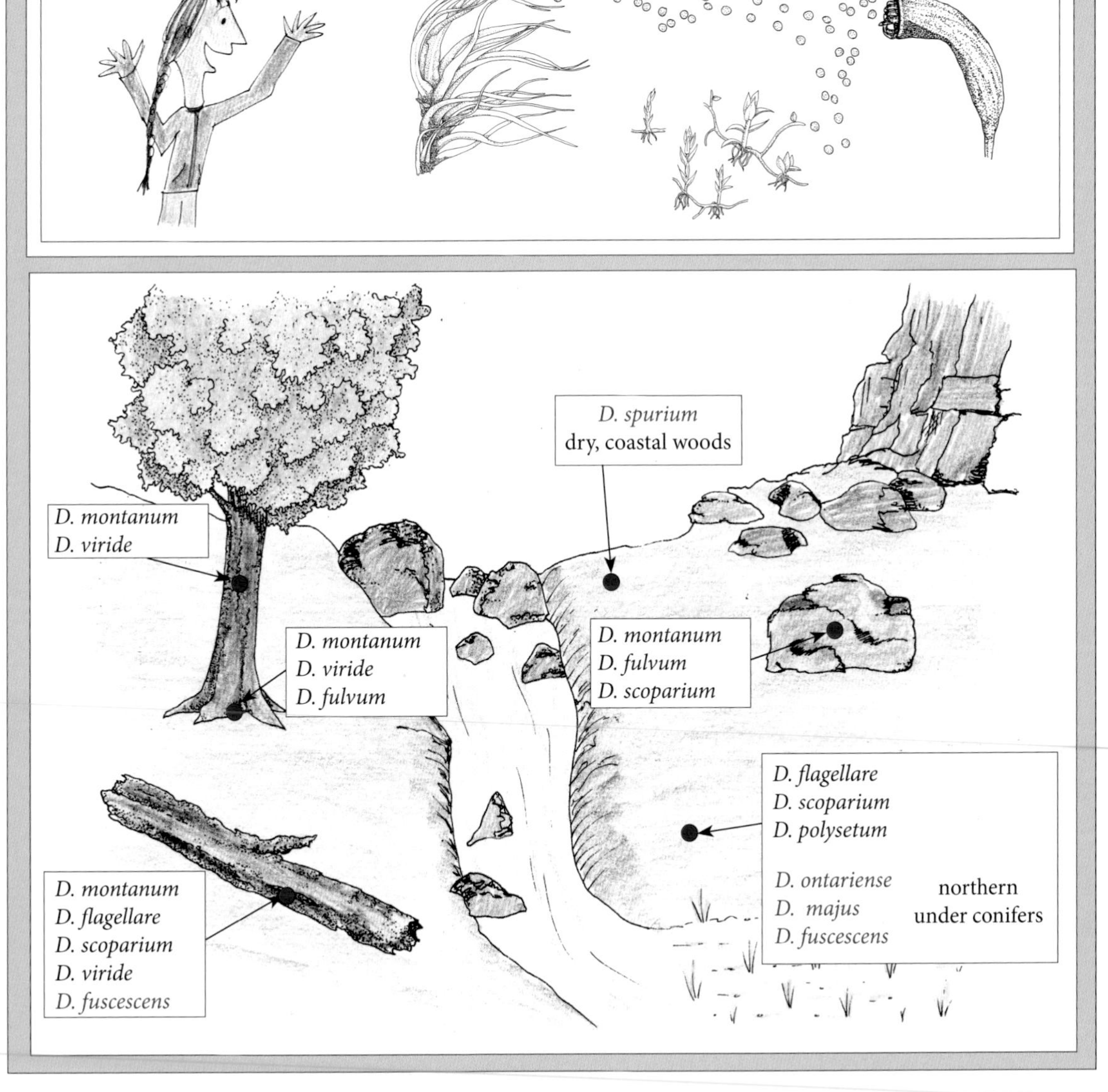

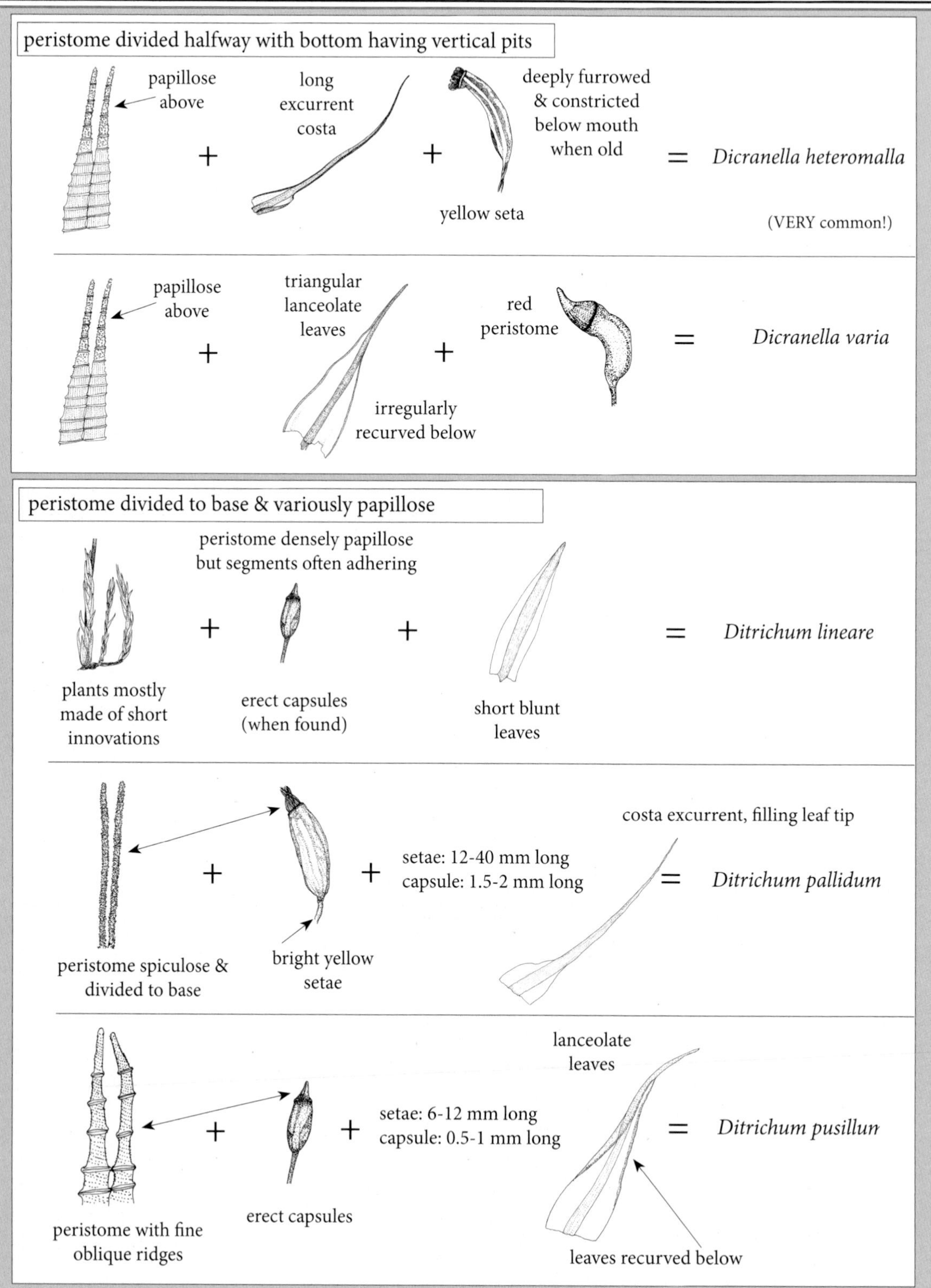
peristome divided halfway with bottom having vertical pits
papillose above
+
long excurrent costa
+
deeply furrowed & constricted below mouth when old
yellow seta
= Dicranella heteromalla
(VERY common!)
papillose above
+
triangular lanceolate leaves
irregularly recurved below
+
red peristome
= Dicranella varia
peristome divided to base & variously papillose
plants mostly made of short innovations
+
peristome densely papillose but segments often adhering
erect capsules (when found)
+
short blunt leaves
= Ditrichum lineare
peristome spiculose & divided to base
+
bright yellow setae
+
setae: 12-40 mm long
capsule: 1.5-2 mm long
costa excurrent, filling leaf tip
= Ditrichum pallidum
peristome with fine oblique ridges
+
erect capsules
+
setae: 6-12 mm long
capsule: 0.5-1 mm long
lanceolate leaves
leaves recurved below
= Ditrichum pusillun

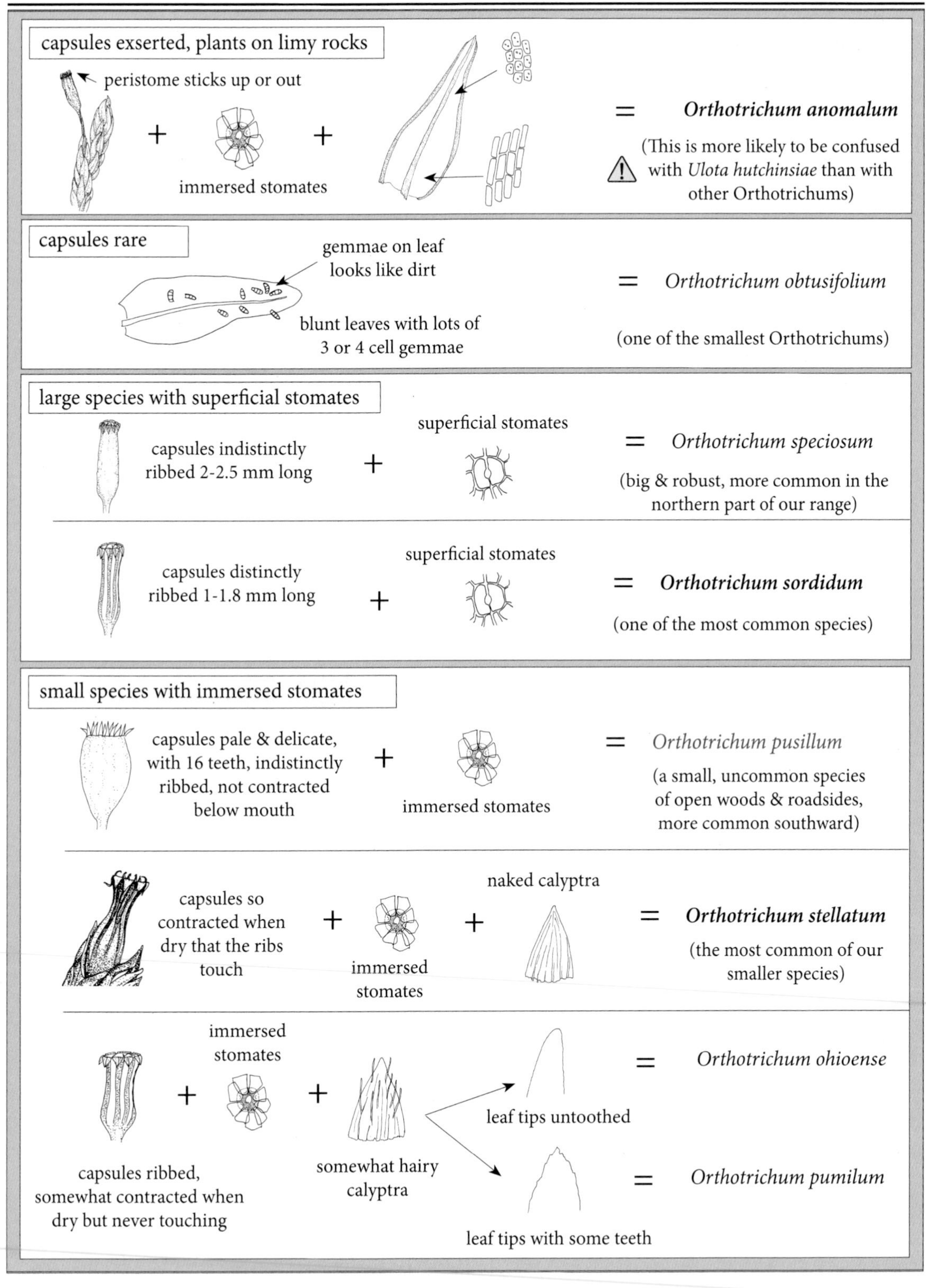
capsules exserted, plants on limy rocks
peristome sticks up or out
+
immersed stomates
+
= Orthotrichum anomalum
(This is more likely to be confused with Ulota hutchinsiae than with other Orthotrichums)
capsules rare
gemmae on leaf looks like dirt
blunt leaves with lots of 3 or 4 cell gemmae
= Orthotrichum obtusifolium
(one of the smallest Orthotrichums)
large species with superficial stomates
capsules indistinctly ribbed 2-2.5 mm long
+
superficial stomates
= Orthotrichum speciosum
(big & robust, more common in the northern part of our range)
capsules distinctly ribbed 1-1.8 mm long
+
superficial stomates
= Orthotrichum sordidum
(one of the most common species)
small species with immersed stomates
capsules pale & delicate, with 16 teeth, indistinctly ribbed, not contracted below mouth
+
immersed stomates
= Orthotrichum pusillum
(a small, uncommon species of open woods & roadsides, more common southward)
capsules so contracted when dry that the ribs touch
+
immersed stomates
+
naked calyptra
= Orthotrichum stellatum
(the most common of our smaller species)
immersed stomates
+
+
somewhat hairy calyptra
capsules ribbed, somewhat contracted when dry but never touching
leaf tips untoothed
= Orthotrichum ohioense
leaf tips with some teeth
= Orthotrichum pumilum

FISSIDENS

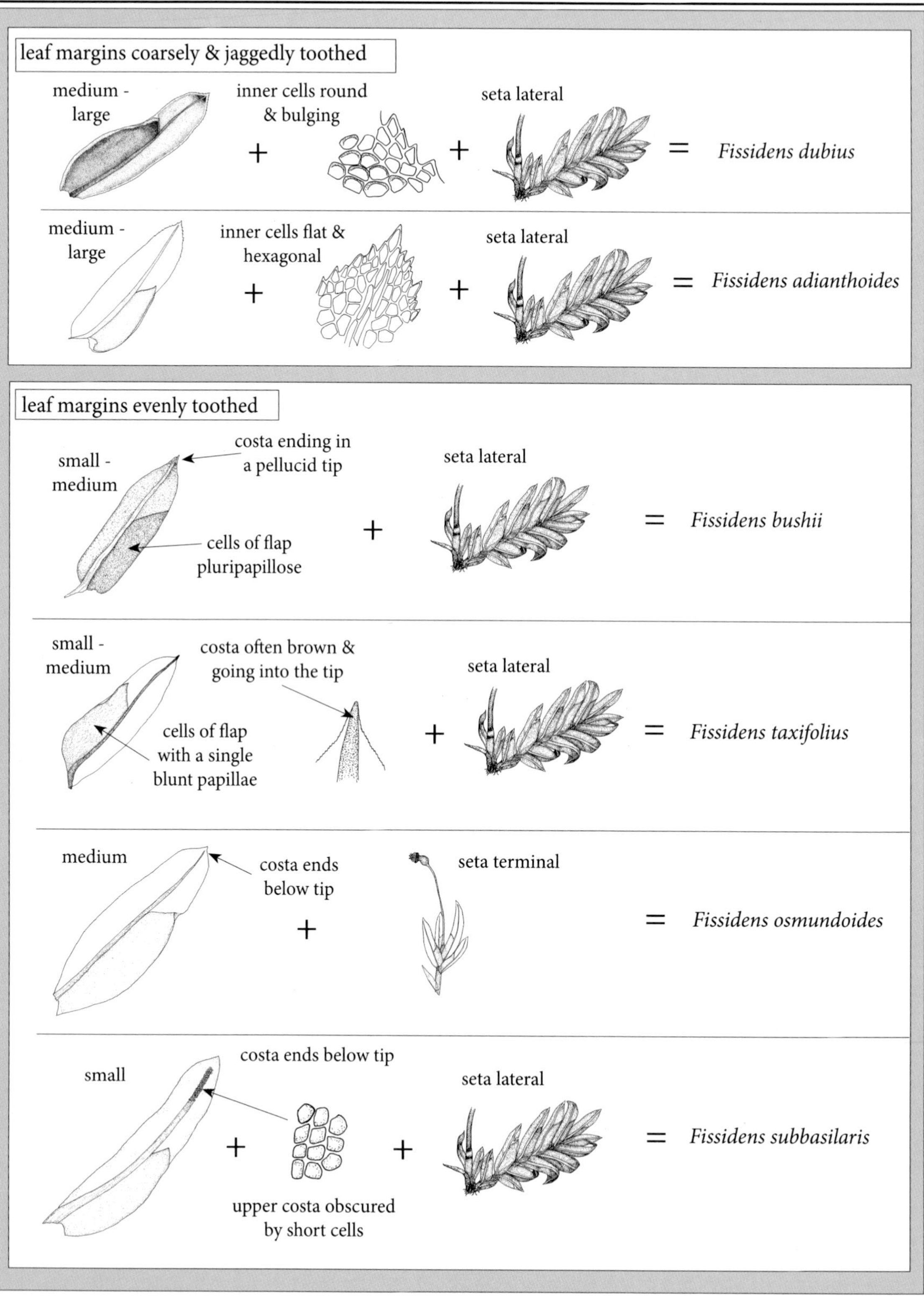

leaf margins coarsely & jaggedly toothed
medium - large
inner cells round & bulging
+
+
seta lateral
=
Fissidens dubius
medium - large
inner cells flat & hexagonal
+
+
seta lateral
=
Fissidens adianthoides
leaf margins evenly toothed
small - medium
costa ending in a pellucid tip
cells of flap pluripapillose
+
seta lateral
=
Fissidens bushii
small - medium
costa often brown & going into the tip
cells of flap with a single blunt papillae
+
seta lateral
=
Fissidens taxifolius
medium
costa ends below tip
+
seta terminal
=
Fissidens osmundoides
small
costa ends below tip
+
upper costa obscured by short cells
+
seta lateral
=
Fissidens subbasilaris

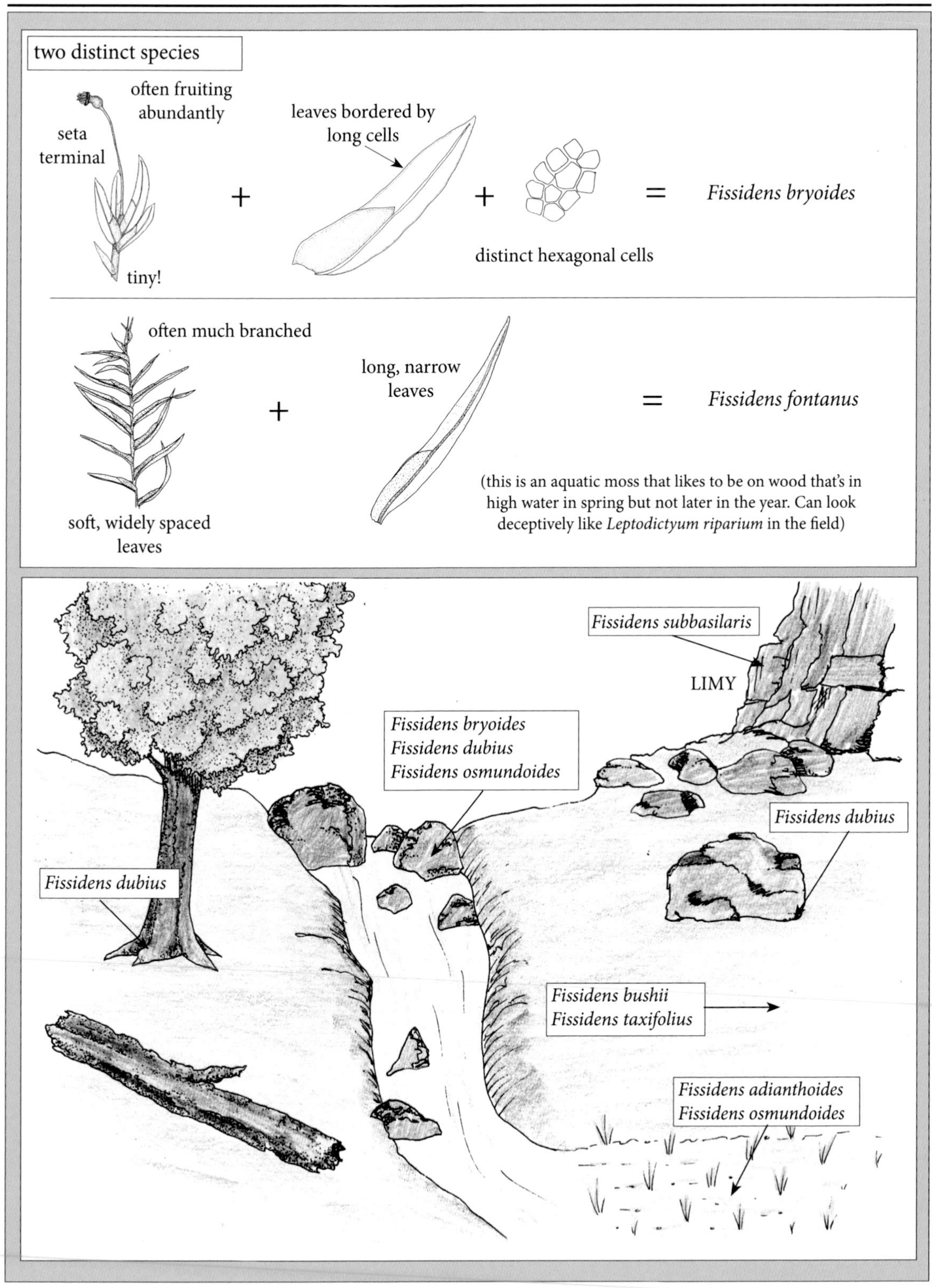
two distinct species
often fruiting abundantly
seta terminal
tiny!
leaves bordered by long cells
distinct hexagonal cells
Fissidens bryoides
often much branched
soft, widely spaced leaves
long, narrow leaves
Fissidens fontanus
(this is an aquatic moss that likes to be on wood that's in high water in spring but not later in the year. Can look deceptively like Leptodictyum riparium in the field)
Fissidens subbasilaris
LIMY
Fissidens bryoides
Fissidens dubius
Fissidens osmundoides
Fissidens dubius
Fissidens dubius
Fissidens bushii
Fissidens taxifolius
Fissidens adianthoides
Fissidens osmundoides

Brachythecium

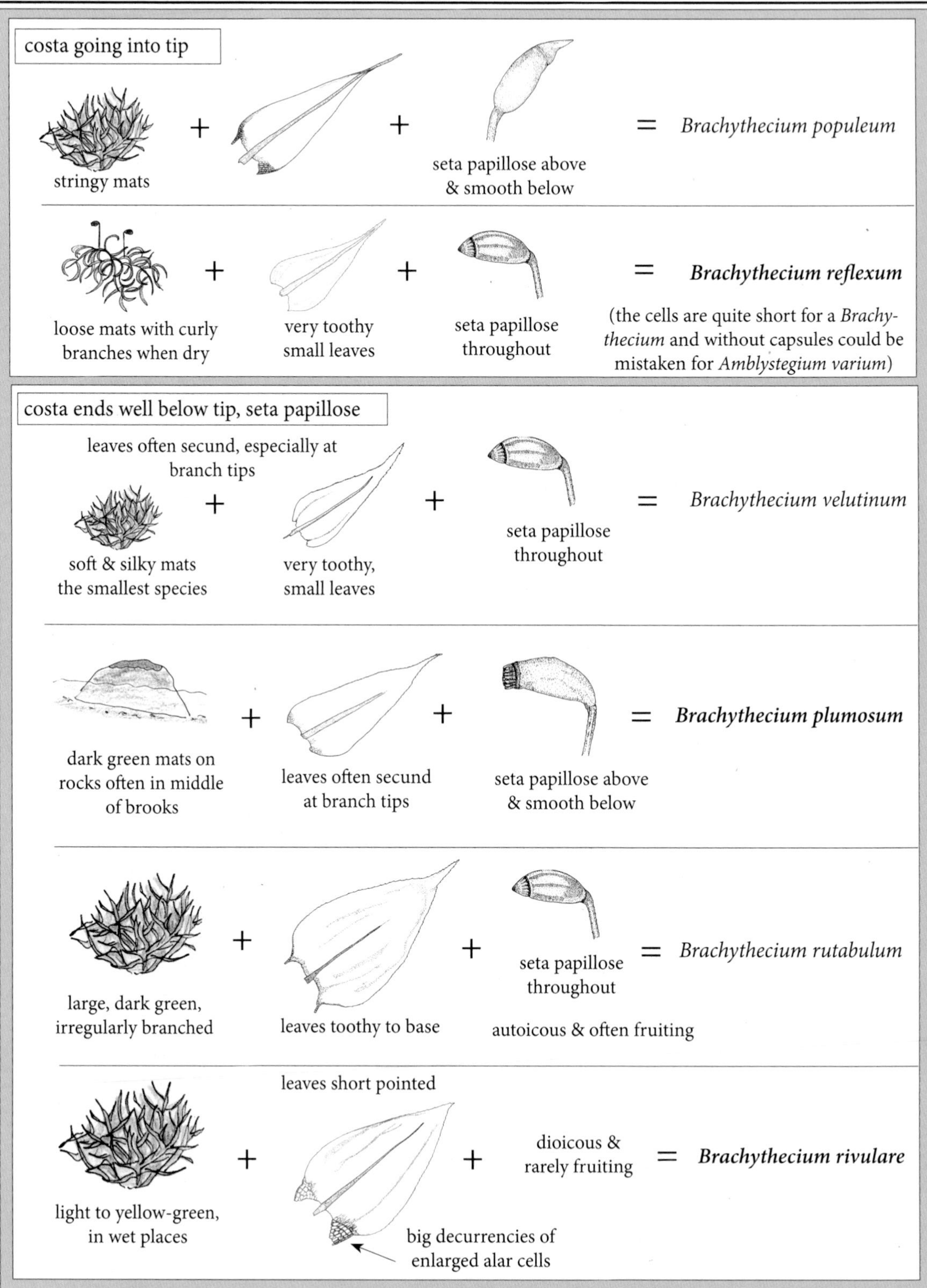

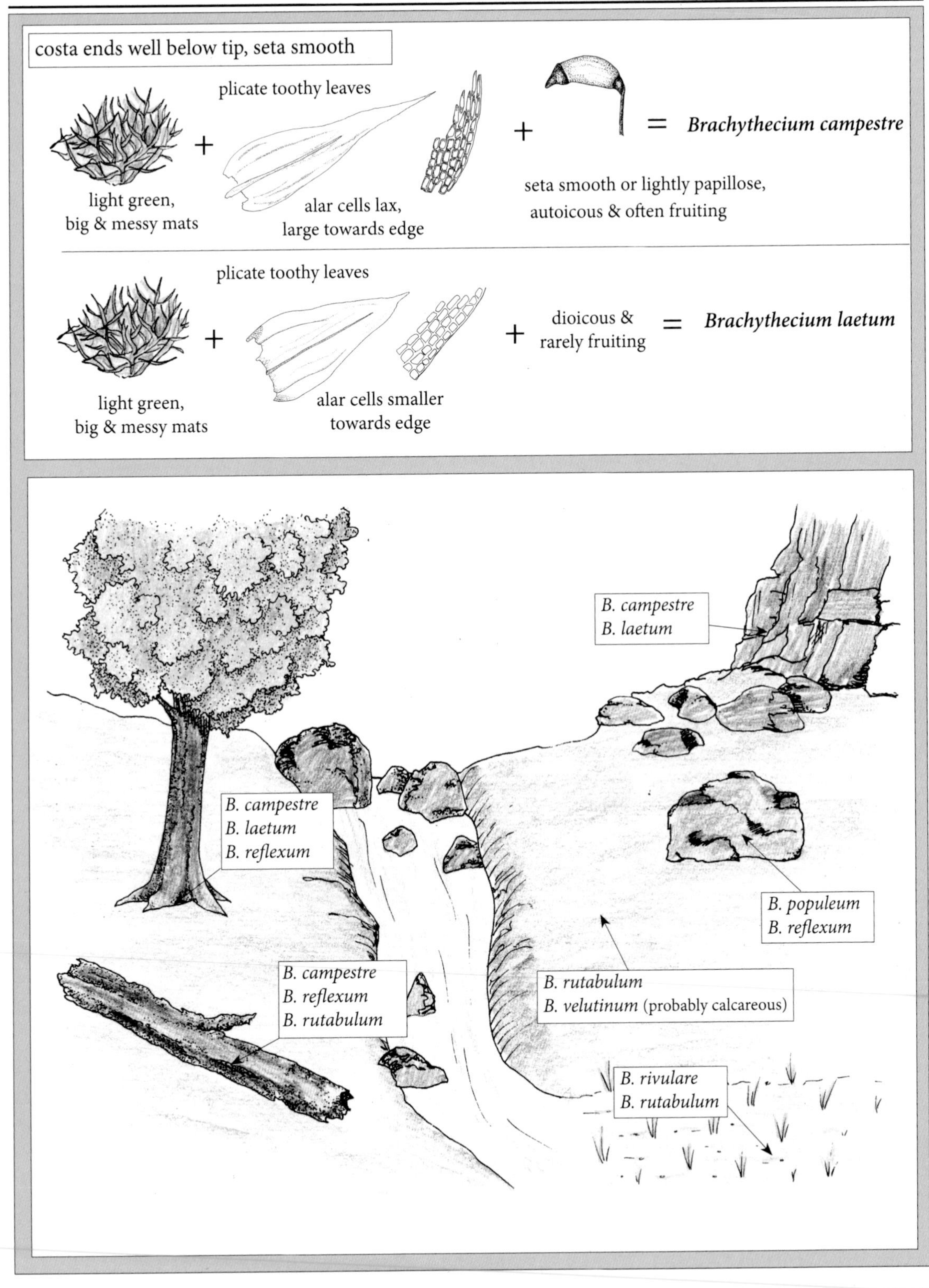
costa ends well below tip, seta smooth
plicate toothy leaves
light green,
big & messy mats
alar cells lax,
large towards edge
seta smooth or lightly papillose,
autoicous & often fruiting
= Brachythecium campestre
plicate toothy leaves
light green,
big & messy mats
alar cells smaller
towards edge
dioicous &
rarely fruiting
= Brachythecium laetum
B. campestre
B. laetum
B. campestre
B. laetum
B. reflexum
B. populeum
B. reflexum
B. campestre
B. reflexum
B. rutabulum
B. rutabulum
B. velutinum (probably calcareous)
B. rivulare
B. rutabulum

AMBLYSTEGIUMS

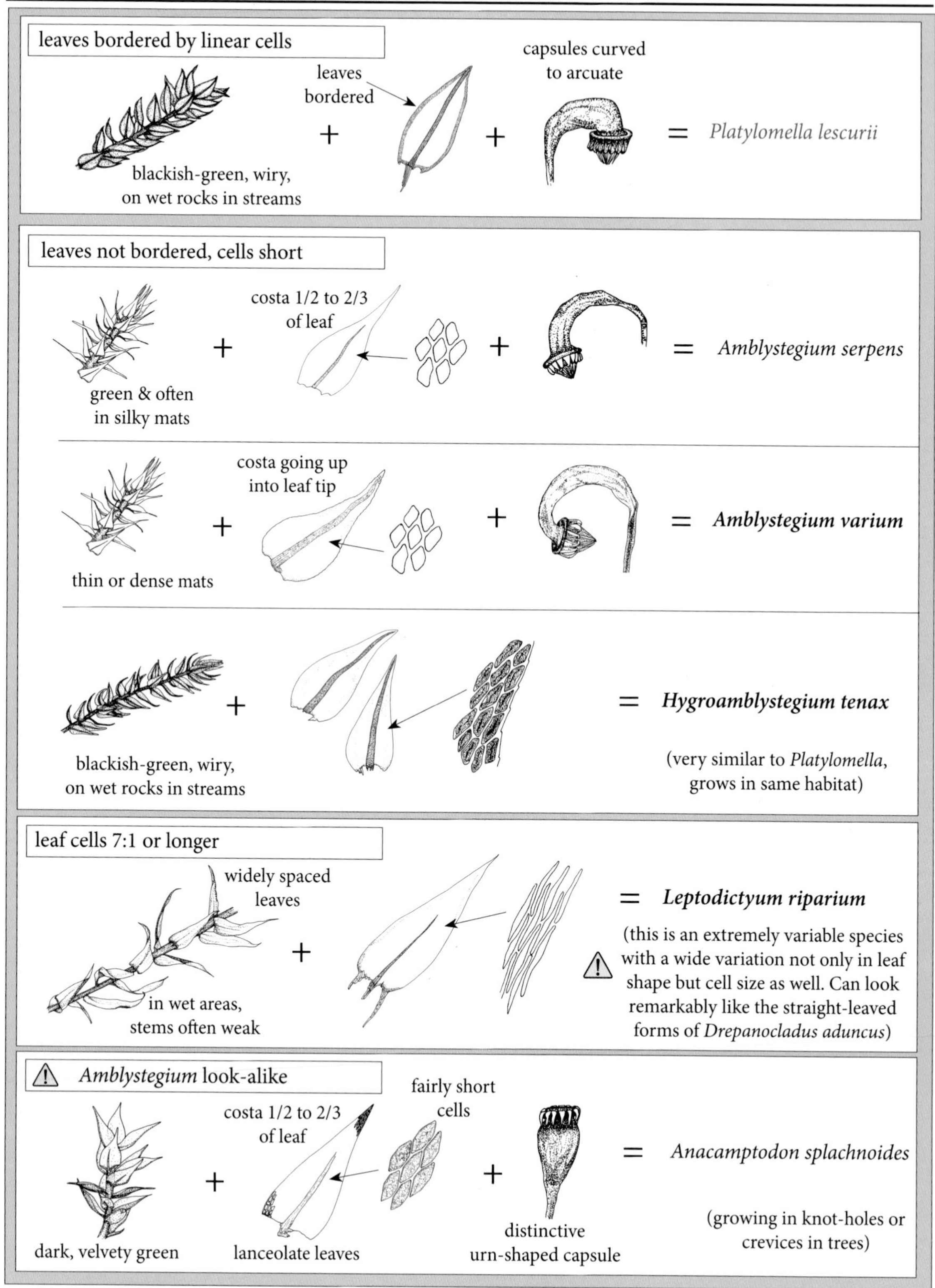

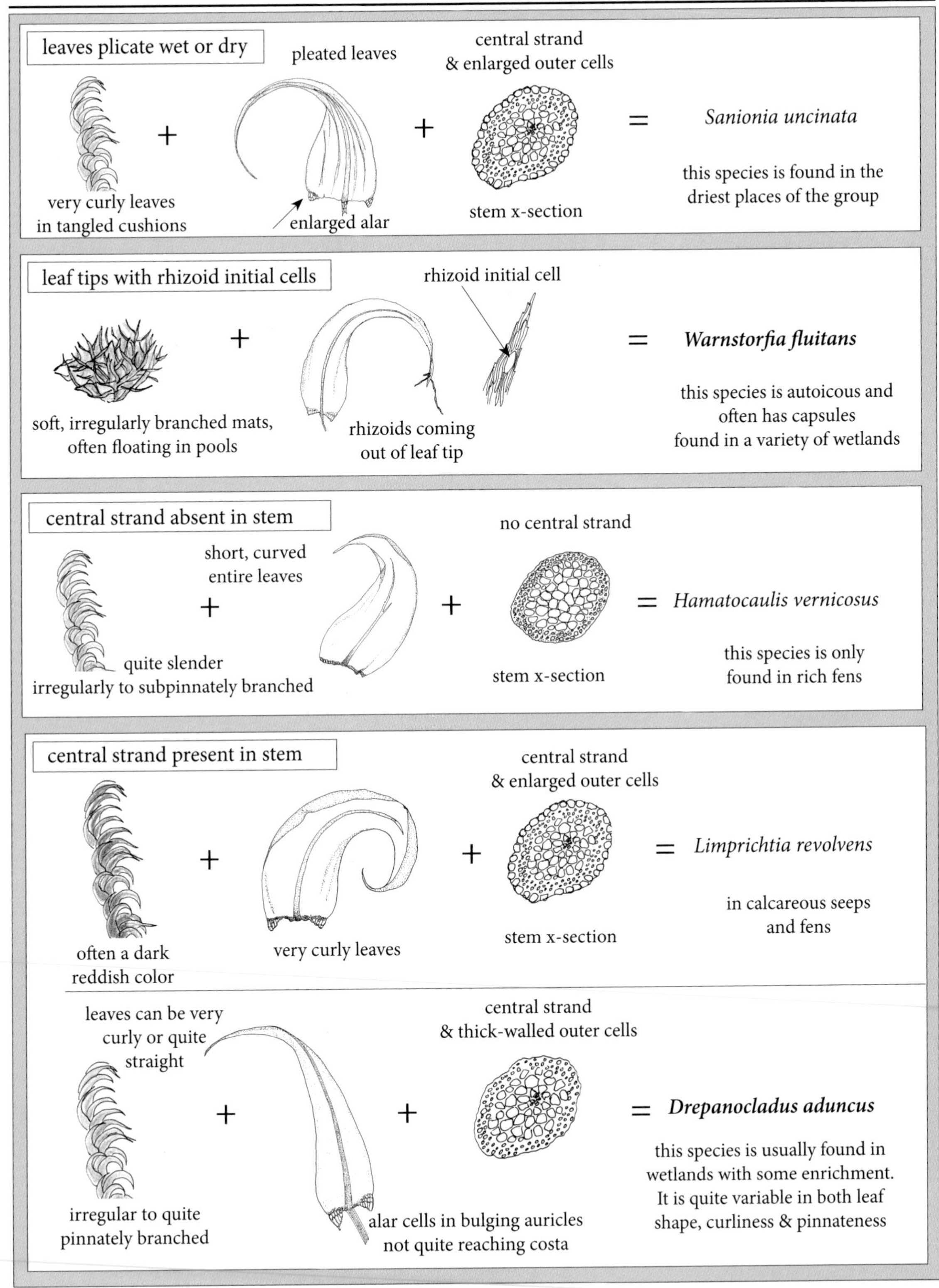
leaves plicate wet or dry
pleated leaves
central strand
& enlarged outer cells
+
+
=
Sanionia uncinata
this species is found in the
driest places of the group
very curly leaves
in tangled cushions
enlarged alar
stem x-section
leaf tips with rhizoid initial cells
rhizoid initial cell
+
=
Warnstorfia fluitans
this species is autoicous and
often has capsules
found in a variety of wetlands
soft, irregularly branched mats,
often floating in pools
rhizoids coming
out of leaf tip
central strand absent in stem
no central strand
short, curved
entire leaves
+
+
=
Hamatocaulis vernicosus
this species is only
found in rich fens
quite slender
irregularly to subpinnately branched
stem x-section
central strand present in stem
central strand
& enlarged outer cells
+
+
=
Limprichtia revolvens
in calcareous seeps
and fens
often a dark
reddish color
very curly leaves
stem x-section
leaves can be very
curly or quite
straight
central strand
& thick-walled outer cells
+
+
=
Drepanocladus aduncus
this species is usually found in
wetlands with some enrichment.
It is quite variable in both leaf
shape, curliness & pinnateness
irregular to quite
pinnately branched
alar cells in bulging auricles
not quite reaching costa

Hygrohypnums

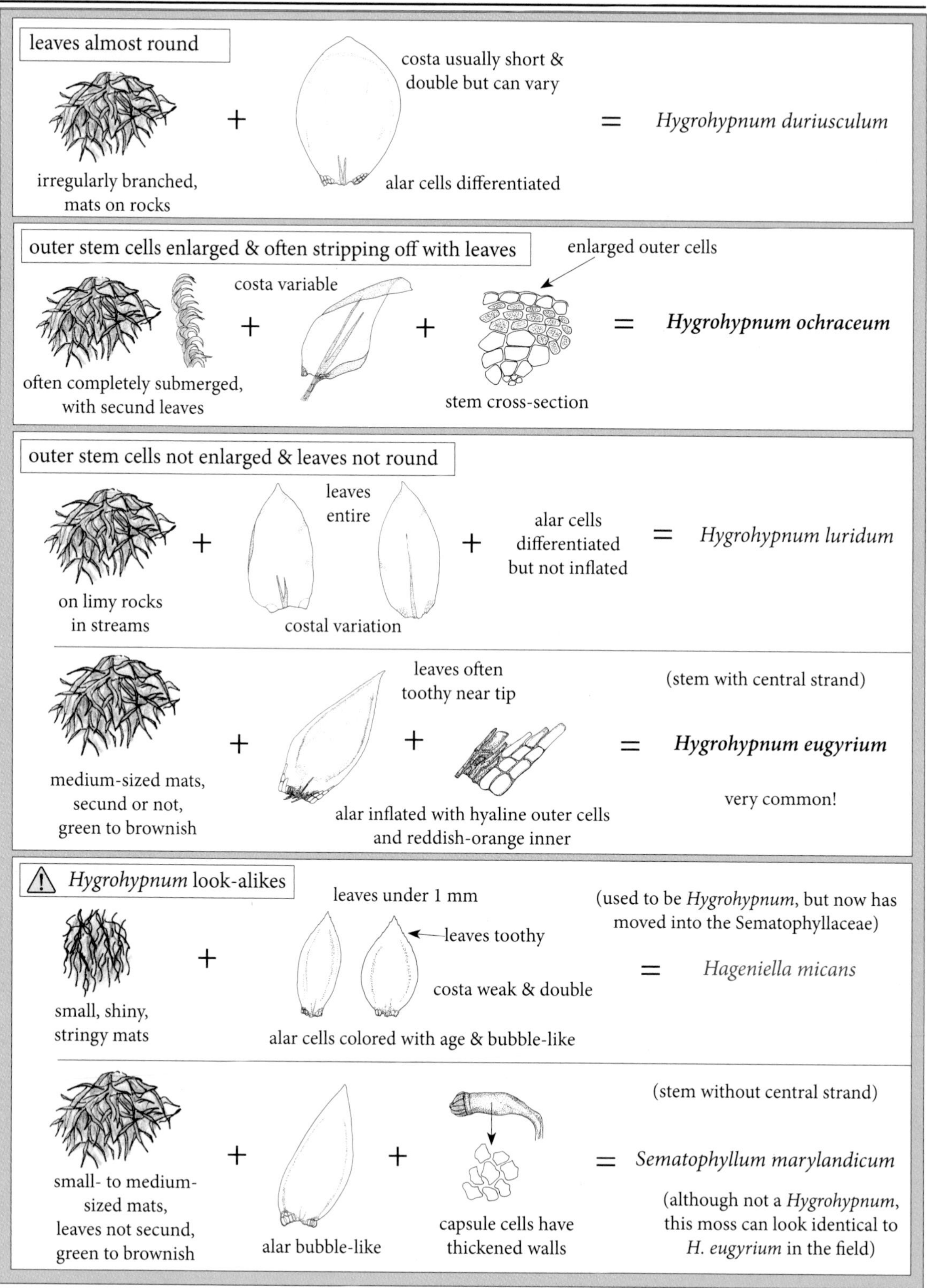

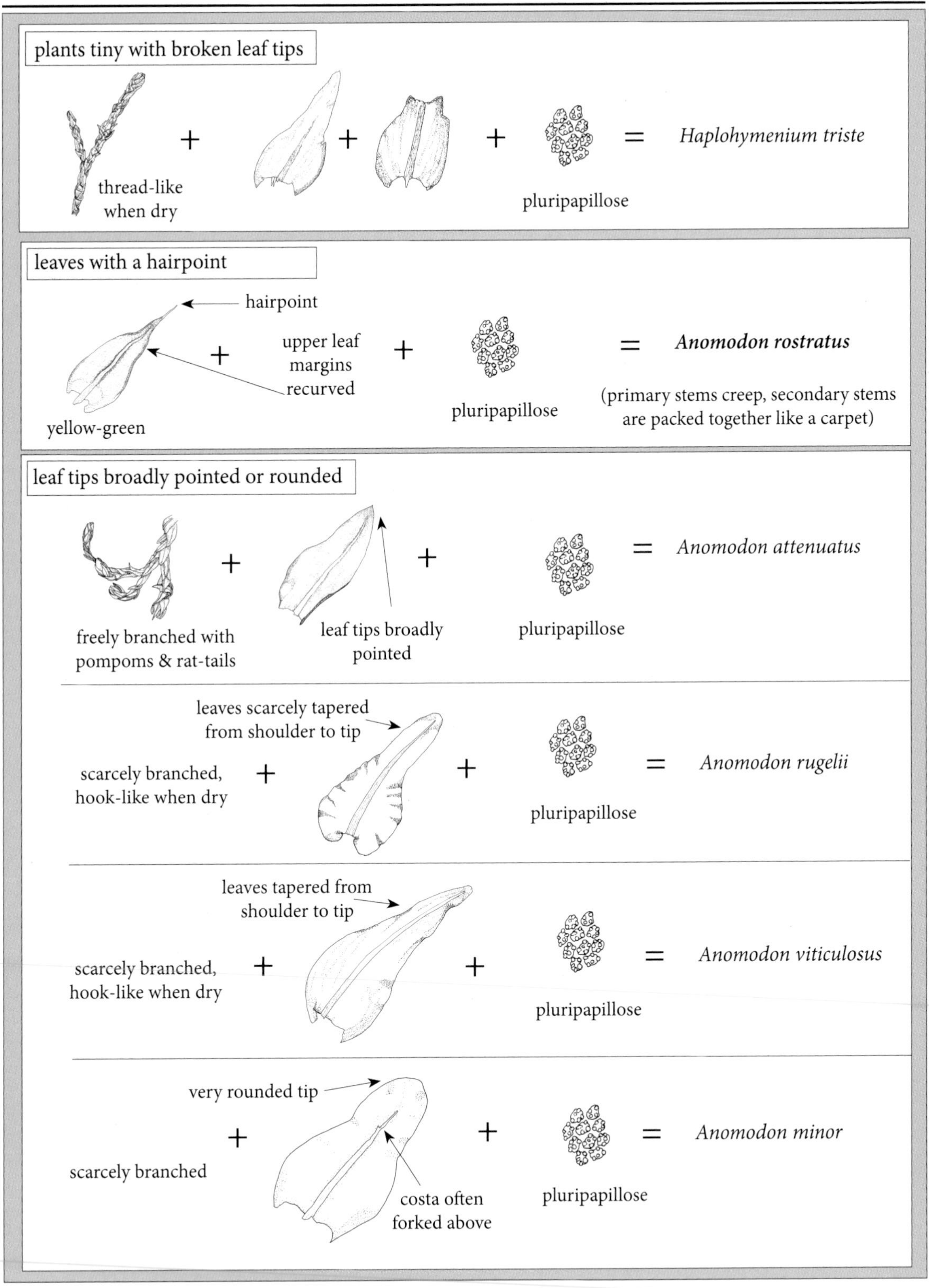
plants tiny with broken leaf tips
thread-like when dry
+
+
+
pluripapillose
=
Haplohymenium triste
leaves with a hairpoint
hairpoint
yellow-green
+
upper leaf margins recurved
+
pluripapillose
=
Anomodon rostratus
(primary stems creep, secondary stems are packed together like a carpet)
leaf tips broadly pointed or rounded
freely branched with pompoms & rat-tails
+
leaf tips broadly pointed
+
pluripapillose
=
Anomodon attenuatus
scarcely branched, hook-like when dry
+
leaves scarcely tapered from shoulder to tip
+
pluripapillose
=
Anomodon rugelii
scarcely branched, hook-like when dry
+
leaves tapered from shoulder to tip
+
pluripapillose
=
Anomodon viticulosus
scarcely branched
+
very rounded tip
costa often forked above
+
pluripapillose
=
Anomodon minor

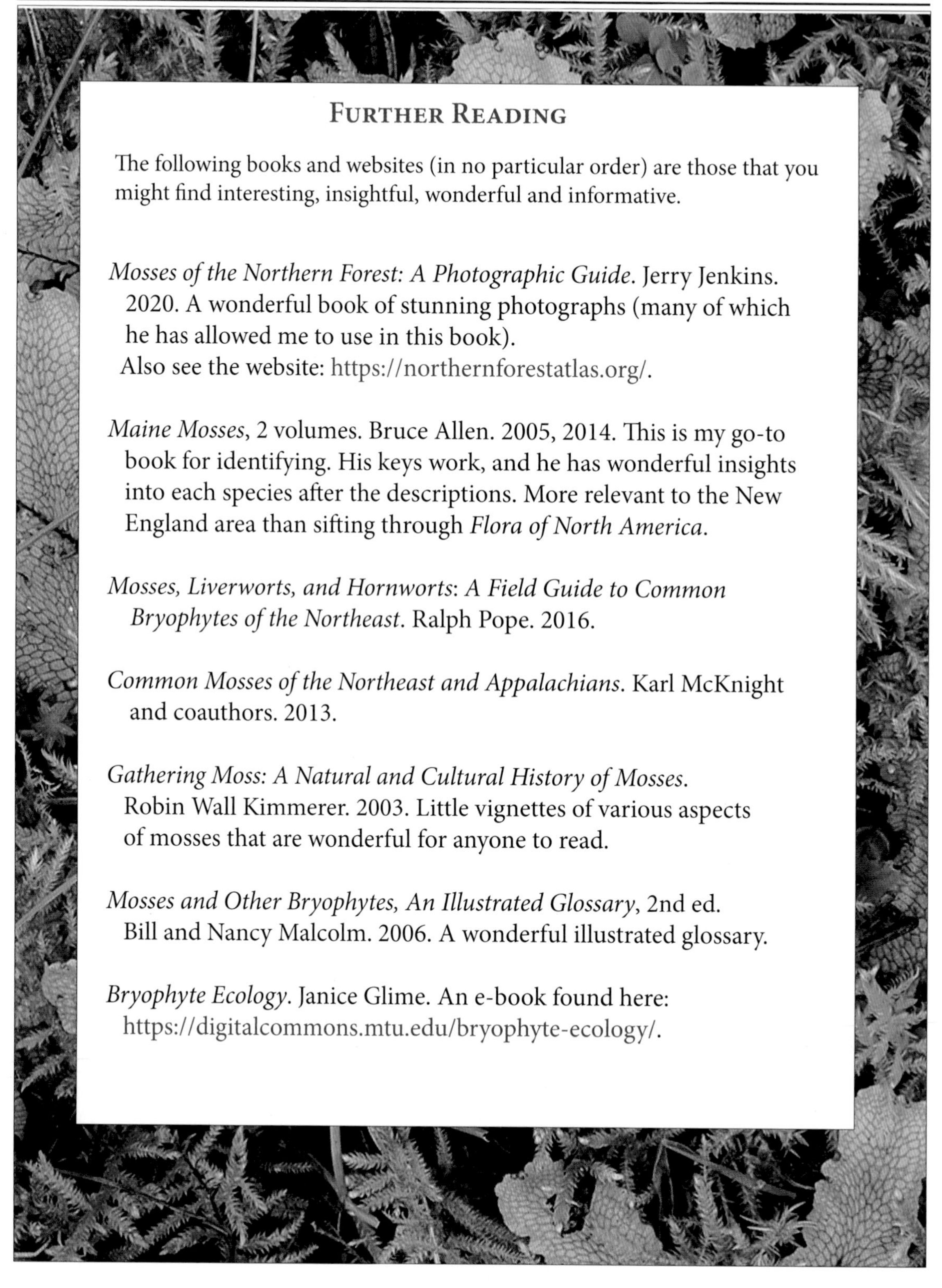

FURTHER READING

The following books and websites (in no particular order) are those that you might find interesting, insightful, wonderful and informative.

Mosses of the Northern Forest: A Photographic Guide. Jerry Jenkins. 2020. A wonderful book of stunning photographs (many of which he has allowed me to use in this book).
Also see the website: https://northernforestatlas.org/.

Maine Mosses, 2 volumes. Bruce Allen. 2005, 2014. This is my go-to book for identifying. His keys work, and he has wonderful insights into each species after the descriptions. More relevant to the New England area than sifting through *Flora of North America*.

Mosses, Liverworts, and Hornworts: A Field Guide to Common Bryophytes of the Northeast. Ralph Pope. 2016.

Common Mosses of the Northeast and Appalachians. Karl McKnight and coauthors. 2013.

Gathering Moss: A Natural and Cultural History of Mosses. Robin Wall Kimmerer. 2003. Little vignettes of various aspects of mosses that are wonderful for anyone to read.

Mosses and Other Bryophytes, An Illustrated Glossary, 2nd ed. Bill and Nancy Malcolm. 2006. A wonderful illustrated glossary.

Bryophyte Ecology. Janice Glime. An e-book found here: https://digitalcommons.mtu.edu/bryophyte-ecology/.

A

Abietinella
- *abietina* *96, 97, 99, 102, 103, 124, 127, 132, 134*

Amblystegium
- *serpens* *85, 90, 198*
- *varium* *37, 41, 45, 57, 60, 96, 128, 132, 196, 198*

Amblystegium Group *198*

Amphidium *136*
- *lapponicum* *125, 130, 133, 139, 141*
- *mougeotii* *109, 115, 125, 130, 139, 141, 143*

Anacamptodon
- *splachnoides* *27, 31, 33, 198*

Andreaea
- *rothii* *107, 109, 115*
- *rupestris* *109, 115, 119*

Aneura
- *pinguis* *165, 169*

Animalcules 78

Anomodon *201*
- *attenuatus* *24, 25, 28, 31, 33, 36, 37, 42, 46, 96, 124, 128, 132, 134, 139, 142, 201*
- *minor* *201*
- *rostratus* *96, 124, 128, 132, 134, 201*
- *rugelii* *37, 42, 46, 128, 201*
- tristis. See *Haplohymenium triste*
- *viticulosis* *128, 132, 134, 201*

Aphanorrhegma
- *serratum* *82, 88, 91*

Arrhenopterum
- heterostichum. See *Aulacomnium heterostichum*

Asterella
- *tenella* *100*

Atrichum *184*
- *altecristatum* *83, 149, 184*
- *angustatum* *80, 81, 83, 89, 149, 184*
- *crispulum* *83, 149, 184*
- *crispum* *146, 148, 156, 184*
- *undulatum* *83, 89, 149, 156, 184*

Aulacomnium
- *heterostichum* *68, 72, 75*
- *palustre* *12, 107, 146, 149, 156, 158, 162*

B

Barbilophozia
- *barbata* *14, 71*

Barbula
- fallax. See *Didymodon fallax*
- reflexa. See *Didymodon ferrugineus*
- *unguiculata* *80, 81, 83, 87, 92*

Bartramia
- *pomiformis* *109, 116, 119, 126*

Bazzania
- *trilobata* *14, 43, 50, 58, 65, 71, 154*

Blasia
- *pusilla* *14, 86, 93*

Blindia
- *acuta* *138, 141*

Brachytheciastrum
- velutinum. See *Brachythecium velutinum*

Brachythecium *42, 69, 196*
- *campestre* *37, 42, 46, 50, 56, 60, 69, 73, 113, 118, 129, 197*
- *curtum* *69*
- *laetum* *37, 42, 46, 50, 56, 113, 118, 129, 132, 134, 197*
- oxycladon. See *Brachythecium laetum*
- *plumosum* *136, 138, 142, 196*
- *populeum* *113, 118, 196*
- *reflexum* *37, 41, 46, 49, 57, 60, 113, 118, 196*
- *rivulare* *56, 139, 142, 151, 157, 196*
- *rutabulum* *56, 60, 69, 73, 196*
- salebrosum. See *Brachythecium campestre*
- *velutinum* *85, 90, 93, 196*

Brotherella
- *recurvans* *40, 45, 49, 57, 60, 62, 70, 74*

Bruchia
- *flexuosa* *82, 88*

Bryhnia
- *graminicolor* *96, 97, 100, 102*
- *novae-angliae* *56, 107, 139, 142, 144, 146, 151, 157*

Bryoandersonia
- *illecebra* *96, 99, 102, 103*

Bryoerythrophyllum
- *recurvirostrum* *125, 131, 133*

Bryum *186*
- *argenteum* *12, 82, 87, 91, 186*
- *caespiticium* *84, 87, 186*
- *capillare* *66, 72, 76, 186*
- *lisae* var. *cuspidatum* *81, 84, 87, 186*
- *pseudotriquetrum* *146, 148, 155, 159, 162, 186*

Bucklandiella
- venusta. See *Racomitrium venustum*

Buxbaumia
- *aphylla* *81, 82, 91*
- *minakatae* *52, 54, 61*

C

Callicladium
- *haldanianum* *36, 37, 42, 45, 50, 52, 55, 56, 60, 112, 117*

Calliergon
- *cordifolium* *150, 157, 159, 164*
- *giganteum* *150, 164, 167, 169*
- trifarium. See *Pseudocalliergon trifarium*

Calliergonella
- *cuspidata* *162, 163, 166, 168*
- lindbergii. See *Hypnum lindbergii*

Calypogeia
- *fissa* *86*
- *muelleriana* *81, 86, 93*

Campyliadelphus
- chrysophyllus. See *Campylium chrysophyllum*

Campylium
- *chrysophyllum* *85, 90, 93*
- *stellatum* *164, 166, 168*

Campylostelium
- *saxicola* *106, 108, 115*

Ceratodon
- *purpureus* *81, 83, 87, 92, 108, 116*

Claopodium
- rostratum. See *Anomodon rostratus*

Climacium
- *americanum* *150, 157*
- *dendroides* *150, 157, 159*

Codriophorus
acicularis. See *Racomitrium aciculare*
aduncoides. See *Racomitrium aduncoides*
Conocephalum
conicum. *See C. salebrosum*
salebrosum 14, 100, 104, 129, 140, 154, 160
Cratoneuron
filicinum 162, 164, 167
Ctenidium
molluscum 96, 100, 102, 104
subrectifolium. See *C. molluscum*
Cyrto-hypnum
pygmaeum. *See Pelekium pygmaeum*

D

Dichodontium
pellucidum 136, 139, 141, 144
Dicranella 192
heteromalla 81, 84, 88, 92, 192
varia 98, 101, 192
Dicranum 190
flagellare 12, 52, 54, 59, 66, 72, 190
fulvum 38, 44, 47, 106, 107, 109, 116, 119, 190
fuscescens 191
majus 191
montanum 24, 25, 26, 30, 32, 36, 37, 38, 44, 47, 54, 59, 107, 108, 116, 190
ontariense 64, 191
polysetum 64, 65, 66, 72, 75, 190
scoparium 54, 59, 64, 65, 66, 72, 75, 107, 110, 116, 190
spurium 190
viride 12, 24, 26, 30, 32, 37, 38, 44, 47, 54, 59, 190
Didymodon
fallax 97, 98, 101, 125, 130
ferrugineus 98, 101, 103, 125, 130
Diphyscium
foliosum 81, 82, 87, 91
Diplophyllum
apiculatum 14, 81, 86
Distichium
capillaceum 98, 101, 103
Ditrichum 192
lineare 84, 87, 192
pallidum 84, 88, 192
pusillum 84, 88, 192
Drepanocladus
aduncus 152, 162, 165, 167, 168, 198, 199
fluitans. *See Warnstorfia fluitans*
revolvens. See *Limprichtia revolvens*
uncinatus. See *Sanionia uncinata*
vernicosus. See *Hamatocaulis vernicosus*
Drepanocladus Group *199*
Drummondia
prorepens 27, 30, 32

E

Elodium
blandowii. See *Helodium blandowii*
paludosum. See *Helodium paludosum*
Encalypta
procera 126, 130, 133
Entodon
cladorrhizans 55, 60, 62, 112, 117, 121
seductrix 39, 45, 48, 113, 117, 122
Ephemerum
crassinervium 82, 88
Eurhynchiastrum
pulchellum. See *Eurhynchium pulchellum*
Eurhynchium
hians 96, 97, 100, 102
pulchellum 69, 73
riparioides. See *Torrentaria riparioides*

F

Fissidens 39, 194
adianthoides 137, 141, 143, 165, 166, 194
bryoides 137, 141, 143, 195
bushii 85, 90, 93, 99, 194
dubius 39, 44, 47, 99, 112, 116, 122, 126, 131, 137, 141, 165, 194
fontanus 195
osmundoides 126, 194
subbasilaris 126, 131, 133, 194
taxifolius 85, 88, 96, 97, 99, 101, 194
Fontinalis 136
antipyretica 137, 142, 144
Forsstroemia
trichomitria 129, 132, 134
Frullania
bolanderi 29
brittoniae 29
eboracensis 14, 24, 25, 29, 34
Funaria
hygrometrica 82, 87, 91

G

Gemmabryum
caespiticium. See *Bryum caespiticium*
Goblin's Gold *94*
Grimmia
pilifera 115, 120
Gymnostomum 139
aeruginosum 125, 130
recurvirostrum. See *Hymenostylium recurvirostrum*

H

Hageniella
micans 137, 200
Hamatocaulis
vernicosus 165, 167, 199
Haplohymenium
triste 12, 25, 28, 31, 201
Hedwigia
ciliata 107, 111, 120
Helodium
blandowii 164, 167, 169
paludosum 151, 157
Herzogiella
striatella 66, 74, 81, 107, 110, 117
Homalia
trichomanoides 39, 46, 47, 106, 112, 118, 121
Homomallium
adnatum 96, 113, 116, 128, 131
Hygroamblystegium
tenax 136, 139, 142, 144, 198
varium. See *Amblystegium varium*

Hygrohypnum 200
duriusculum 138, 200
eugyrium 136, 137, 142, 144, 200
luridum 137, 200
ochraceum 137, 142, 200
Hylocomiastrum
umbratum. See *Hylocomium umbratum*
Hylocomium
splendens 64, 65, 70, 74, 77, 151, 157
umbratum 151, 156
Hymenostylium
recurvirostrum 125
Hyophila
involuta 136, 138, 141, 143
Hypnum
cupressiforme 111, 116
imponens 37, 40, 45, 49, 52, 56, 60, 64, 65, 70, 74, 81, 111, 117, 121
lindbergii 69, 74, 146, 152, 156, 158, 162
pallescens 24, 25, 27, 31, 33, 36, 37, 40, 45, 49, 57, 60, 107, 111, 117, 121

I

Isopterygiopsis
muelleriana 112, 116, 121
Isopterygium
distichaceum. See *Pseudotaxiphyllum distichaceum*
elegans. See *Pseudotaxiphyllum elegans*

J

Jamesoniella
autumnalis. See *Syzygiella autumnalis*
Jubula
pennsylvanica 136, 140

L

Lepidozia
reptans 71, 77
Leptobryum
pyriforme 84, 88
Leptodictyum
riparium 146, 152, 157, 195, 198
Leskea 31, 41, 45
gracilescens 28, 33, 41
obscura 28, 41
polycarpa 28, 41
Leskeella
nervosa 12, 27, 31, 37, 41, 45
Leucobryum 81
glaucum 55, 59, 61, 66, 72, 107, 110, 115
Leucodon
andrewsianus 24, 25, 28, 31, 34
sciuroides. See *L. andrewsianus*
Lewinskya
sordida. See *Orthotrichum sordidum*
Limprichtia
revolvens 162, 165, 167, 199
Lophocolea
heterophylla 37, 43, 52, 58, 62

M

Mannia
fragrans 100
Marsupella
emarginata 140
Meesia
triquetra 162, 163, 166, 168
Metzgeria
conjugata 114, 122, 129
Mnium
appalachianum. See *Rhizomnium appalachianum*
ciliare. See *Plagiomnium ciliare*
cuspidatum. See *Plagiomnium cuspidatum*
ellipticum. See *Plagiomnium ellipticum*
hornum 68, 73, 76, 149, 156, 189
marginatum 127, 131, 189
punctatum. See *Rhizomnium punctatum*
spinulosum 68, 73, 76, 189
stellare 68, 73, 75, 188
thomsonii 127, 131, 189
Mnium Group 188
Myurella
julacea 128, 131, 134
sibirica 128, 131, 134

N

Neckera
pennata 24, 25, 28, 31, 34
Nowellia
curvifolia 52, 58, 62
Nyholmiella
obtusifolia. See *Orthotrichum obtusifolium*

O

Odontoschisma
denudatum 52
Oncophorus
wahlenbergii 52, 54, 59, 61
Orthodicranum
flagellare. See *Dicranum flagellare*
fulvum. See *Dicranum fulvum*
montanum. See *Dicranum montanum*
viride. See *Dicranum viride*
Orthotrichum 24, 30, 193
anomalum 109, 125, 130, 193
obtusifolium 12, 25, 26, 32, 193
ohioense 26, 193
pumilum 193
pusillum 193
sordidum 25, 26, 32, 193
speciosum 26, 193
stellatum 25, 26, 32, 193
strangulatum 125, 130
Oxyrrynchium
hians. See *Eurhynchium hians*
Oxystegus
tenuirostris. See *Trichostomum tenuirostre*

P

Pallavicinia
lyellii 14, 154, 160
Paludella
squarrosa 162, 163, 166, 168
Paraleucobryum
longifolium 110, 115, 119
Pelekium
pygmaeum 128,132
Pellia
epiphylla 154, 160

Philonotis
fontana *146, 148, 156, 158, 163*
marchica *163, 166*
Physcomitrium
pyriforme *80, 82, 88*
serratum. See *Aphanorrhegma*
Plagiochila
porelloides *14, 140*
Plagiomnium
ciliare *55, 59, 62, 68, 72, 149, 155, 188*
cuspidatum *37, 38, 44, 47, 55, 59, 68, 72, 188*
ellipticum *149, 155, 188*
Plagiopus
oederianus *126, 131, 133*
Plagiothecium *39, 45, 106*
cavifolium *12, 107, 112, 113, 117, 121, 137, 142*
denticulatum *39, 48, 85, 90, 107, 112, 117, 137*
laetum *37, 39, 85, 90, 112, 117, 137*
latebricola *52, 55, 60*
Platydictya *128*
confervoides *96, 131*
Platygyrium
repens *24, 25, 27, 31, 33, 36, 37, 41, 45, 57, 60*
Platyhypnidium
riparioides. See *Torrentaria riparioides*
Platylomella
lescurii *139, 142, 144, 198*
Pleuridium
subulatum *82, 87, 91*
Pleurozium
schreberi *64, 65, 70, 74, 77, 106, 107, 163*
Pogonatum
alpinum. See *Polytrichastrum alpinum*
dentatum *81, 86, 89, 92*
pensilvanicum *80, 82, 89, 91*
urnigerum *86, 89*
Pohlia *187*
andalusica *148, 187*
annotina *12, 146, 148, 155, 158, 187*
bulbifera *146, 148, 155, 158, 187*
cruda *107, 127, 131, 187*
nutans *81, 84, 88, 187*
wahlenbergii *148, 155, 163, 187*
Polytrichastrum *185*
alpinum *67, 73, 110, 116, 120, 185*
formosum. See *Polytrichum formosum*
longisetum. See *Polytrichum longisetum*
ohioense. See *Polytrichum ohioense*
pallidisetum. See *Polytrichum pallidisetum*
Polytrichum *81, 185*
commune *65, 73, 76, 81, 86, 89, 107, 110, 116, 120, 185*
formosum *67*
juniperinum *67, 73, 77, 81, 86, 89, 110, 115, 120, 185*
longisetum *67*
ohioense *67, 73, 185*
pallidisetum *67, 73, 106, 185*
piliferum *81, 86, 89, 92, 110, 115, 120, 185*
Porella
pinnata *136, 140*
platyphylla *14, 24, 25, 29, 34, 37, 43, 50, 114, 122, 129*
Preissia
quadrata *100, 129*
Pseudobryum
cinclidioides *150, 155, 158, 188*
Pseudocalliergon
trifarium *162, 163, 167*
Pseudotaxiphyllum *12*
distichaceum *85, 90, 93, 112, 117*
elegans *85, 90, 112, 117*
Pterigynandrum
filiforme *41, 45, 48, 107, 113, 117*
Ptilidium
ciliare *65, 71, 77*
pulcherrimum *25, 29, 34, 37, 43, 50, 58, 107, 114, 122*
Ptilium
crista-castrensis *56, 60, 62, 64, 65, 70, 74*
Ptychostomum
creberrimum. See *Bryum lisae* var. *cuspidatum*
pseudotriquetrum. See *Bryum pseudotriquetrum*
Pylaisia *31*
condensata. See *P. selwynii*
intricata *27*
polyantha *27*
selwynii *25, 27, 33*
Pylaisiadelpha
recurvans. See *Brotherella recurvans*
Pylaisiella
intricata. See *Pylaisia intricata*
polyantha. See *Pylaisia polyantha*
selwynii. See *Pylaisia selwynii*

R

Racomitrium
aciculare *138, 141, 143*
aduncoides *138, 141, 143*
heterostichum. See *R. venustum*
venustum *111, 118*
Radula
complanata *24, 29, 34, 37, 43, 50, 114*
Rauiella
scita *37, 40, 46, 49*
Reboulia
hemisphaerica *100, 129*
Rhabdoweisia
crispata *108, 116*
Rhizomnium
appalachianum *150, 155, 158, 189*
magnifolium *150*
punctatum *138, 141, 143, 189*
Rhodobryum
ontariense *67, 72, 76, 127, 131*
Rhynchostegium
serrulatum *39, 46, 48*
Rhytidiadelphus
squarrosus *151, 156, 159*
subpinnatus. See *Rhytidiadelphus squarrosus*
triquetrus *65, 69, 74, 75, 150, 157, 159*

Rhytidium
rugosum *96, 97, 99, 102, 104, 127, 132*
Riccardia
latifrons *52*
multifida *14, 154*
palmata *52*
Rosulabryum
capillare. See *Bryum capillare*
flaccidum. See *Bryum capillare*
Rotifers *78*

S

Saelania
glaucescens *97, 98, 101, 103, 125, 131*
Sanionia
uncinata *199*
Scapania
nemorea *107, 140*
Schistidium
apocarpum *109, 111, 115, 119, 138*
rivulare *138, 142*
Schistostega
pennata *94*
Schwetschkeopsis
fabronia *128, 131, 134*
Sciaromium
lescurii. See *Platylomella lescurii*
Sciuro-hypnum
plumosum. See *Brachythecium plumosum*
populeum. See *Brachythecium populeum*
reflexum. See *Brachythecium reflexum*
Scorpidium
revolvens. See *Limprichtia revolvens*
scorpioides *162, 163, 166, 168*
Sematophyllum
demissum *113, 116*
marylandicum *137, 142, 200*
Serpoleskea
confervoides. See *Platydictya confervoides*
Sphagnum *146, 147*
angustifolium *152*
capillifolium *152, 159*
divinum. See *S. magellanicum*
fallax *152*
fimbriatum *153*
flexuosum *152*
girgensohnii *152*
magellanicum *153*
medium. See *S. magellanicum*
palustre *153, 160*
recurvum *152*
recurvum group *152, 160*
rubellum *165*
squarrosum *153, 159*
subsecundum *146, 153*
warnstorfii *165, 169*
wulfianum *153, 159*
Splachnum
ampullaceum *170*
luteum *170*
rubrum *170*
Steerecleus
serrulatus. See *Rhynchostegium serrulatum*
Syntrichia
papillosa *12, 26, 30*
ruralis *126, 130, 133*
Syzygiella
autumnalis *37, 58, 114, 122*

T

Tardigrades *78*
Taxiphyllum
deplanatum *96, 99, 102, 104*
Tetraphis
pellucida *12, 53, 55, 59, 61, 68, 72, 75*
Tetraplodon *170*
angustatus *170*
mnioides *170*
Thamnobryum
alleghaniense *106, 113, 118, 122*
Thelia *39*
asprella *39, 46, 48*
hirtella *39, 46*
Thuidium
delicatulum *37, 40, 46, 49, 52, 53, 56, 60, 70, 73, 99, 111, 118, 121, 127, 139, 142, 150, 157, 162, 164*
pygmaeum. See *Pelekium pygmaeum*
recognitum *56, 70, 96, 97, 99, 102, 104, 111, 127, 132, 150, 164, 167, 169*
scitum. See *Rauiella scita*
Timmia
megapolitana *96, 97, 98, 101, 103*
Tomenthypnum
nitens. See *Tomentypnum nitens*
Tomentypnum
nitens *162, 164, 167, 169*
Torrentaria
riparioides *136, 137, 142*
Tortella
humilis *126, 130, 133*
inclinata *126, 130, 133*
tortuosa *96, 97, 98, 101, 124, 126, 130, 133*
Tortula
papillosa. See *Syntrichia papillosa*
ruralis. See *Syntrichia ruralis*
truncata *82, 87*
Trematodon
ambiguus *84, 87, 92*
longicollis *84, 87, 92*
Trichocolea
tomentella *154, 160*
Trichostomum
tenuirostre *108, 115*

U

Ulota
coarctata *26, 30, 32*
crispa *24, 25, 26, 30, 32, 37, 38, 44*
crispula. See *U. crispa*
hutchinsiae *107, 109, 115, 119, 125, 193*

W

Warnstorfia
fluitans *146, 152, 157, 199*
Water Bears *78*
Weissia
controversa *81, 83, 87, 92, 98, 108, 115*
muhlenbergiana *98, 101*